Effect of IL-10 and anti-TGF-beta antibodies on the morphology of bone marrow stroma cultures
from
Interleukin-10
by
Jan E. DeVries and René de Waal Malefyt

Molecular
Biology
Intelligence
Unit

The *Drosophila* Genome Map: A Practical Guide

Daniel L. Hartl

Elena R. Lozovskaya

Harvard University
Cambridge, Massachusetts, U.S.A.

R.G. Landes Company
Austin

Molecular Biology Intelligence Unit

THE *DROSOPHILA* GENOME MAP: A PRACTICAL GUIDE

R.G. LANDES COMPANY
Austin, Texas, U.S.A.

Submitted: December1994
Published: March 1995

Please address all inquiries to the Publisher:
R.G. Landes Company, 909 Pine Street, Georgetown, Texas, U.S.A. 78626
or
P.O. Box 4858, Austin, Texas, U.S.A. 78765
Phone: 512/ 863 7762; FAX: 512/ 863 0081

U.S. and Canada ISBN 1-57059-101-6

International ISBN 3-540-59094-3

Library of Congress Cataloging-in-Publication Data

Hartl, Daniel L.
The drosophila genome map: a practical guide / Daniel L. Hartl, Elena R. Lozovskaya.
p. cm. — (Molecular biology intelligence unit)
Includes bibliographical references and index.
ISBN 1-57059-101-6 (hardcover : alk. paper)
1. Drosophila — Genome mapping. I. Lozovskaya, Elena R. II. Title III. Series.
[DNLM: 1. Drosophila melanogaster — genetics. 2. Chromosome Mapping.
QX 505 H331d 1995]
QL537.D76h383 1995
595.77'4—dc20
DNLM/DLC for Library of Congress

PUBLISHER'S NOTE

R.G. Landes Company publishes five book series: *Medical Intelligence Unit, Molecular Biology Intelligence Unit, Neuroscience Intelligence Unit, Tissue Engineering Intelligence Unit* and *Biotechnology Intelligence Unit.* The authors of our books are acknowledged leaders in their fields and the topics are unique. Almost without exception, no other similar books exist on these topics.

Our goal is to publish books in important and rapidly changing areas of medicine for sophisticated researchers and clinicians. To achieve this goal, we have accelerated our publishing program to conform to the fast pace in which information grows in biomedical science. Most of our books are published within 90 to 120 days of receipt of the manuscript. We would like to thank our readers for their continuing interest and welcome any comments or suggestions they may have for future books.

Deborah Muir Molsberry
Publications Director
R.G. Landes Company

CONTENTS

PREFACE

Our approach to mapping the *Drosophila* genome is based on large fragments of genomic DNA cloned in various vectors. The rationale was designed to take maximum advantage of the giant polytene chromosomes present in the salivary-gland nuclei of third instar larvae. The initial strategy was to assign locations to individual clones by in situ hybridization with the polytene chromosomes. Following the in situ mapping of one or two genome equivalents of clones chosen at random, overlapping clones could be identified and gaps in the map filled by screening the libraries with probes from a number of sources. One advantage of this approach is that it has established the framework of the physical map relatively rapidly. Equally as important, even while the initial in situ mapping was in progress, the project generated research materials in the form of large-fragment clones with identified locations in the genome that were useful to many *Drosophila* researchers.

Completion of the physical map of the *Drosophila* genome is one of the priorities of the first phase of the *Drosophila* Genome Center described below. This objective is well on the way to being achieved. In its present stage of completion, the map already includes such an extensive set of P1 and YAC clones that we have decided to summarize the available materials and provide protocols for their use. Researchers wishing electronic access to the data and other relevant information may contact ftp.bio.indiana.edu in the directory flybase or else contact flybase@morgan.harvard.edu. The information in these sources also includes a list of laboratories throughout the world from whom P1 clones may be requested.

This book was written in preparation for a practical course on the *Drosophila* genome that we presented at the Gulbenkian Doctoral Program in Biology and Medicine in Portugal. This enlightened program was initiated by the Advanced Study Institute of the Gulbenkian Foundation, and it is also funded by the Portuguese Secretaries of State for Higher Education and for Science and Technology. We are grateful to the Directors of the Program for the opportunity to establish contact with Portuguese students and research laboratories. We are also deeply indebted to Claudio E. Sunkel and Ms. Paula Almeida Coelho of the University of Porto for their hard work in the preparation and teaching of the course.

The physical map of *Drosophila* based on large-fragment clones would not have been possible without the continuing support of the National Center for Human Genome Research (NCHGR). From the beginning of the Human Genome Project, the fruit fly, *Drosophila melanogaster,* was among the select group of species included as a model organism deserving of special attention.[1]

Initial support for the construction of the large-fragment libraries and for the in situ mapping of large numbers of clones was provided by the NCHGR through an investigator-initiated research grant (RO1) to D. L. Hartl and Ian Duncan (Washington University). At the present time, the project is supported by the NCHGR through the *Drosophila* Genome Center, headed by Gerald M. Rubin (University of California, Berkeley). The *Drosophila* Genome Center was established by a consortium of investigators including G. M. Rubin (University of California, Berkeley), Daniel L. Hartl and Elena R. Lozovskaya (Harvard), Allan C. Spradling (Carnegie Institution of Washington), and Michael J. Palazzolo and Christopher H. Martin (Lawrence Berkeley Laboratories). Partial support for the *Drosophila* genome project also comes from the Howard Hughes Medical Institute through its support of the Rubin and Spradling laboratories. In addition to its previous support, the NCHGR has announced a 1993-1998 5-year plan that includes a specific commitment to provide funding adequate to sequence the entire *Drosophila* genome.[2] Large parts of the Bithorax and Antennapedia regions have already been sequenced by a group of investigators headed by M. J. Palazzolo and C. H. Martin at the Lawrence Berkeley Laboratories, and the NCHGR has provided support to the Lawrence Berkeley group for the large-scale genomic sequencing of a region of approximately two million base pairs around the gene for alcohol dehydrogenase *(Adh)*.

We also want to thank our many coworkers whose participation made the work described in this book possible and often pleasurable. The acknowledgments are in alphabetical order. We are especially indebted to the following people who worked on the project in our own laboratory: James W. Ajioka, Lara Brilla, John P. Carulli, O. N. Danilevskaya, Michael Devine, Dan Garza, Dana M. Hartl, Theodore J. Hartl, Ann Honeycutt, Robert W. Jones, Margaret A. Lackey, Allan R. Lohe, Dmitry I. Nurminsky, Dmitri Petrov, Jennifer Schutzman, David A. Smoller, Michael Takamura, Patricia Tsai, Ann E. C. Vellek, and Jie Wei.

We also wish to thank the following collaborators: Michael Ashburner, David T. Burke, Haini Cai, Georges F. Carle, Ian W. Duncan, Jorg D. Hoheisel, Daniel H. Johnson, Fotis Kafatos, Paula Kiefel, Hans Lehrach, Andrew J. Link, Cristos Louis, Christopher H. Martin, C. Mayeda, H. Ochman, Maynard V. Olson, Gerald M. Rubin, Michael J. Palazzolo, Stanley A. Sawyer, and Joyce Yee.

References

1. Watson JD. The human genome project: Past, present, and future. Science 1990; 248:44-49.
2. Collins F, Galas D. A new five-year plan for the U.S. human genome project. Science 1993; 262:43-49.

CHAPTER 1

ORIGIN

The *Drosophila* genome project summarized in this book was begun in the year 1987. The origins of the project trace to St. Louis and the efforts of three enthusiastic graduate students—Dan Garza, David T. Burke, and Georges F. Carle. Garza was a graduate student in the Hartl laboratory. He was big, tall, strong fellow—built more like the stereotype of a stevedore than that of a scientist—full of ideas, energy, and enthusiasm for research, particularly in genetics. As an undergraduate at Stony Brook, he had worked part-time in New York in the demolition end of a construction company and once, by mistake, had dismantled the better part of the walls in an office complex in the few hours between visits by the foreman. He worked equally hard in graduate school under what one hopes was better guidance. Burke, a graduate student of Maynard V. Olson, followed instructions better than Garza but was also very hardworking as well as something of a perfectionist. Carle, another of Olson's students, was a slim, swarthy Frenchman with the soul and education of an engineer.

The Olson and Hartl laboratories were adjacent, and the three students worked side by side. They also talked a lot and swapped ideas. Especially at night, they brainstormed in the small departmental coffee room. Nearly every morning, evidence of the previous night's discussion remained on the blackboard in elaborate diagrams of vectors and genetic experiments and also on the stove in the form of an empty and unwashed coffee pot. Daytime, they would gather at intervals to exchange new ideas or suggest alternative approaches to what they had discussed the night before. The interests of the three were perfectly complementary. Garza loved *Drosophila* genetics and the sheer joy of examining anesthetized flies and pushing them around with an artist's brush under a dissecting microscope. Burke excelled at the manipulation of recombinant DNA and had been constructing specialized vectors for yeast in the hope of being able to clone large fragments of exogenous DNA in the form of artificial chromosomes. Following Schwartz and Cantor,[1] Carle had been trying to develop the experimental conditions to improve the electrophoretic separation of large DNA molecules so that such molecules could be sized accurately and isolated.[2]

By the time the *Drosophila* genome project began in earnest, Burke and Carle were well along in their own research. Working with Carle in the Olson laboratory, Burke had already made a bacterial vector containing a centromeric sequence from a yeast chromosome and telomeric sequences from *Tetrahymena*, and the group had shown that genetically stable yeast artificial chromosomes (YACs) containing large segments of exogenous DNA could be created.[3] Garza had decided to push for a *Drosophila* genome project and had already designed a YAC vector in which the cloning site was flanked by the ends of the transposable element *P* in hope that germline transformation might be successful with large DNA fragments; transformation had already been demonstrated for small DNA fragments flanked by *P*-element ends in plasmids.[4] Within a few weeks, Garza and Burke had made the vector and Garza was preparing large quantities of *Drosophila* DNA to make the first libraries. By that time, James W. Ajioka had joined the Hartl laboratory as a Postdoctoral Fellow and had become an enthusiastic participant in the genome research. Ajioka had been recruited because, in his graduate work at Stony Brook, he had become an expert in interpreting the banding patterns observed in the in situ hybridization with the polytene salivary gland chromosomes. He was originally recruited to do research with *Drosophila* transposable elements,[5] but he quickly became attracted to the *Drosophila* genome project. Because Ajioka was also very good in the benchwork of molecular genetics, his contributions to the *Drosophila* genome research soon became essential.

Even at this early stage of genome research it seemed a good bet that large-fragment DNA cloning and the construction of physical maps of complex genomes might become an important growing point in the development of molecular genetics. There were also early signs that large-scale, automated DNA sequencing might be possible. Thus it happened that the *Drosophila* genome project was begun. Shortly thereafter, Robert W. Jones was recruited as a senior technician because there was too much cytological work for one person to do. Across St. Louis's Forest Park, on the Arts and Sciences campus of Washington University, Ian Duncan took an interest in the project, and so in the Fall of 1987 Hartl and Duncan applied for a grant from the National Institutes of Health to support the research.

The initial funding of the research was provided in September, 1988, by the National Institute of General Medical Sciences. Shortly thereafter, the National Center for Human Genome Research was created and the Human Genome Project became Big Science. The enterprise has been remarkably successful. Like any large human endeavor, the Human Genome Project has had to deal with its share of personal ambition, greed, inability to cooperate, and competition for recognition and scientific prizes. Somehow it managed very skillfully to maneuver the project through and around these obstacles. Thanks to the Human Genome Project and the support of the community of *Drosophila* researchers, the *Drosophila* genome project has thrived. This book has been written in the hope that the materials generated thus far will be maximally useful as a resource in the community of researchers. It presents the physical map of the genome of the fruit fly, *Drosophila melanogaster,* as constructed by in situ hybridization and various methods for identifying overlapping clones in bacteriophage P1 and YAC vectors. Complementary work based on cosmid vectors carried out by a group of European investigators including Michael Ashburner, Fotis Kafatos, Christos Louis, and Robert Saunders[6,7] is not included, primarily owing to limitations of space. The current status of the P1 and YAC physical map is presented with emphasis on resolution, biological content, and integration with the genetic map. The genomic localizations of P1 and YAC clones are tabulated, along with detailed protocols for the genetic manipulation of the clones. Also discussed is the general utility of the set of mapping clones for research in molecular genetics, development, neurobiology, and evolutionary biology.

References

1. Schwartz DC, Cantor CR. Separation of yeast chromosome-sized DNAs by pulsed field gradient gel electrophoresis. Cell 1984; 37:67-75.
2. Carle GF, Olson MV. Separation of chromosomal DNA molecules

from yeast by orthogonal-field-alternation gel electrophoresis. Nucleic Acids Res 1984; 12:5647-5663.

3. Burke DT, Carle GF, Olson MV. Cloning of large segments of exogenous DNA into yeast by means of artificial chromosome vectors. Science 1987; 236:806-812.
4. Rubin GM, Spradling AC. Genetic transformation of *Drosophila* with transposable element vectors. Science 1982; 218:348-353.
5. Ajioka JW, Hartl DL. Population dynamics of transposable elements. In: D. E. Berg and M. M. Howe, eds. *Mobile DNA*. Washington, D.C.: American Society for Microbiology 1989; 939-958.
6. Sidén-Kiamos I, Saunders RDC, Spanos L, Majerus T, Treanear J, Savakis C, Louis C, Glover DM, Ashburner M, Kafatos FC. Towards a physical map of the *Drosophila melanogaster* genome: Mapping of cosmid clones within defined genomic divisions. Nucleic Acids Res 1990; 18:6261-6270.
7. Kafatos FC, Louis C, Savakis C, Glover DM, Ashburner M, Link AJ, Sidén-Kiamos I, Saunders RDC. Integrated maps of the *Drosophila* genome: progress and prospects. Trends Genet 1991; 7:155-161.

CHAPTER 2

P1 AND YAC LIBRARIES

Preparation of libraries containing large fragments of DNA is nontrivial. We usually warn people without prior experience who are planning to make such a library that they should allocate six months or more to the task. Six months is, in any case, approximately the time it took to prepare the YAC and P1 libraries in the Hartl laboratory. The YAC libraries were the first to be prepared and characterized, but it was clear even while the YAC work was progressing that we would have to adopt a "top-down" strategy and eventually need a cloning vector with smaller inserts in order to provide relatively easy access to substrates for DNA sequencing and other uses. At about the same time that the YAC project was begun, a group of European investigators including Michael Ashburner, Fotis Kafatos, Christos Louis, and Robert Saunders had also initiated a *Drosophila* genome project based on cosmid vectors, which have inserts of approximately 30 to 40 kb.[1,2] Their strategy was to isolate particular chromosomal regions by carving out tiny sections of the polytene chromosomes with a micromanipulator. DNA from each microdissected region was amplified by the polymerase chain reaction (PCR), and the resulting material used as a hybridization probe with a cosmid library. Cosmids that hybridized with the probe were isolated and fingerprinted by electrophoresis of small restriction fragments in order to identify overlaps between the inserts in the clones. Finally, the clones were hybridized with the polytene chromosomes in order to confirm their locations in the map.

With two research groups each developing a physical map of *Drosophila*, it would not have been surprising had there been some tension. *Drosophila* is an organism with a relatively small genome, and it could hardly have been parceled out to different research groups, chromosome by chromosome, as the human genome had been. Fortunately, the researchers involved in both *Drosophila* projects understood early on that the projects were complementary and, if successful, would be good for each other as well as for the *Drosophila* community. At breakfast one Friday morning in April of 1989, during the 30th Annual *Drosophila* Research Conference held in New Orleans, Louisiana, the groups formally agreed to share information and materials, and thereafter both projects progressed in parallel. The sharing of materials and information soon embraced a third *Drosophila* genome project, which differed from the others in the use of oligonucleotide hybridization techniques to identify overlapping clones.[3-5]

For those of us working with YACs, it seemed perfectly clear at the time that the next logical vector in a top-down cloning strategy after YACs would have to be cosmids. Or so it seemed until December of 1989 when, at a meeting on Large DNA Fragment Cloning in Houston, Texas, Nat Sternberg gave a presentation on a new cloning system based on bacteriophage P1. One of us (DLH) was fortunate enough to be at the meeting and was tremendously excited about the possibility of applying the P1 cloning system to *Drosophila*. The P1 system was a novel cloning system with many of the advantages of cosmids but with an approximately twofold larger insert size. The potential advantage of the P1 system was that many of the clones could be expected to be large enough to hybridize with multiple adjacent bands in the salivary gland chromosomes and hence could be localized in a manner similar to that used with YACs.

By the time the session in which Sternberg had talked was finished, I (DLH) had stationed myself outside the door of the conference room to wait for Sternberg. After introducing myself, I excitedly explained the role that P1 clones might play in the *Drosophila* genome project. Sternberg briefly described his hopes for the further development of the P1 system and promised his support for the application of the system to *Drosophila*. After the meeting, I returned to St. Louis and immediately tried to interest David A. Smoller, a new Postdoctoral Fellow from Emory University, in the P1 system. Smoller had already learned to prepare high molecular weight DNA from *Drosophila* because he was working with Ajioka in producing new YAC libraries. As an inducement, I told him that a *Drosophila* P1 library could be named after him and would probably become as famous and be used as widely as the celebrated Maniatis phage library. For a few days there was some teasing banter about the nomenclature, and I do not know to what extent the issue was significant to Smoller's motivation. Soon, however, using his experience making YAC libraries, and drawing on the full cooperation of the P1 researchers at DuPont/Merck, he was busy at work constructing the first *Drosophila* P1 library. In recognition of his efforts, the P1 clones listed in this book have official numerical designations preceded by the initials DS.

This book being a practical guide to the *Drosophila* genome, the following sections provide some of the details of library construction. Because P1 clones have taken primacy over YAC clones for the analysis of the *Drosophila* genome, methods for creating the P1 libraries will be discussed first.

THE BACTERIOPHAGE P1 CLONING SYSTEM

Pierce and Sternberg[6] provide detailed descriptions of the P1 cloning system, the preparation of clones, and brief summaries of pertinent features of the P1 life cycle. Readers interested in specific aspects of the cloning system should refer to their paper. Here we will focus on the essential elements of the system. A diagram of the first-generation P1 cloning vector pNS582-tet14 Ad10 is shown in Figure 2.1. The vector is divided into two regions by the pair of *loxP* sites. Each *loxP* site is the following 34 bp sequence containing 13 bp inverted repeats (underlined):

5'-ATAACTTCGTATAATGTATGCTATACGAAGTTAT-3'

The *loxP* sites are targets for the Cre site-specific recombinase[7] and are used for circularization of the vector after DNA insertion, packaging and infection. The region between the *loxP* sites containing the *Sca*I site includes a *pac* site, a plasmid pBR322 replicon, and an 11 kb segment of adenovirus DNA inserted into a unique *Sca*I site in the ampicillin-resistance (*amp-r* or *bla*) gene of pBR322 in such a way as to preserve the unique *Sca*I site in the vector. The *pac* site is a 162 bp sequence that is cleaved approximately in the middle by phage-encoded "pacase" proteins to initiate the headful packaging mechanism of phage P1; the *pac* site contains Dam methylation sequences (5'-GATC-3'), which must be methylated for cleavage. In the cloning vector in Figure 2.1, the *pac* site is so oriented as to direct headful packaging counterclockwise as the illustration is oriented.

On the other side between the *loxP* sites is a region containing a kanamycin-resistance (*kan*) gene from the transposon *Tn*903, the tetracycline-resistance (*tet*) genes from plasmid pBR322, the P1 plasmid replicon (*rep*) and the P1 partition functions that ensure segregation of the single-copy plasmid into daughter cells, and the P1 lytic replicon (*lytic*). The lytic replicon has been altered so that it is regulated by the promoter of the *E. coli* lactose operon (*lac*); expression is repressed in host cells containing the *lacI*^{q} allele, but the lytic replicon can be induced by adding the *lac* operon inducer IPTG (isopropyl 1-thio-β-D-galactopyranoside) to the growth medium prior to harvesting and DNA isolation.

The second-generation P1 vector pAd10 sacBII[8] is diagrammed in Figure 2.2. (The complete sequence is present in GenBank as accession number U09128, courtesy of Bruce A. Roe and collaborators). The pAD10 sacBII vector is identical to the vector in Figure 2.1 except that the region between the *Bam*HI and *Sal*I sites

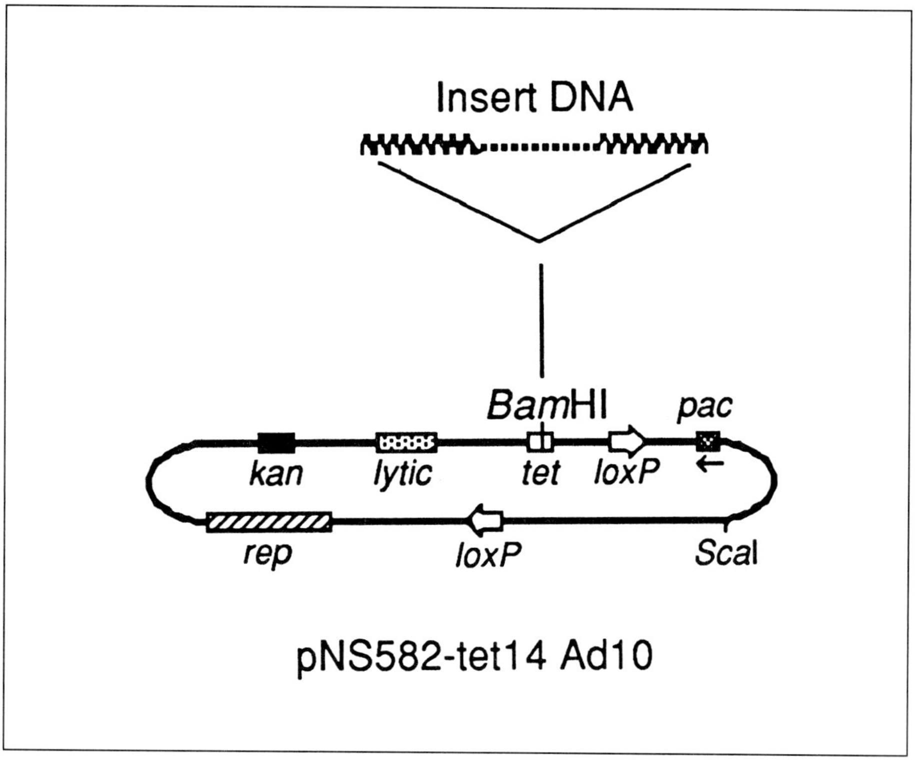

Fig. 2.1. Diagram of the P1 cloning vector pNS582-tet14 Ad10. Data from Sternberg et al, Proc Natl Acad Sci USA 1990; 87:103-107.

in the *tet* gene has been replaced with a 2.0-kb insert (the *sac*BII cassette) containing an *E. coli* promoter that regulates the *sac*BII gene from *Bacillus amyloliquefaciens.* The promoter also includes an overlapping binding site for the c1 suppressor protein of phage P1, so that expression of *sac*BII does not take place in the presence of c1. The *sac*BII gene itself codes for levansucrase, the expression of which kills *E. coli* cells in media containing more than 2% sucrose owing to osmotic shock. Large fragments of DNA inserted into the *Bam*HI cloning site disrupt expression of the *sacB*II gene and thereby allow cells of *E. coli* to survive in medium containing 5% sucrose. The *sac*BII gene in pAd10 sacBII thereby enables a positive selection for vectors with inserts that distance the promoter from the *sac*BII gene. The *Bam*HI cloning site in the vector is itself part of a casette flanked by a bacteriophage T7 promoter and a *Not*I restriction site on the promoter side and by a bacteriophage SP6 promoter and an *Sfi*I restriction site on the *sac*BII side.

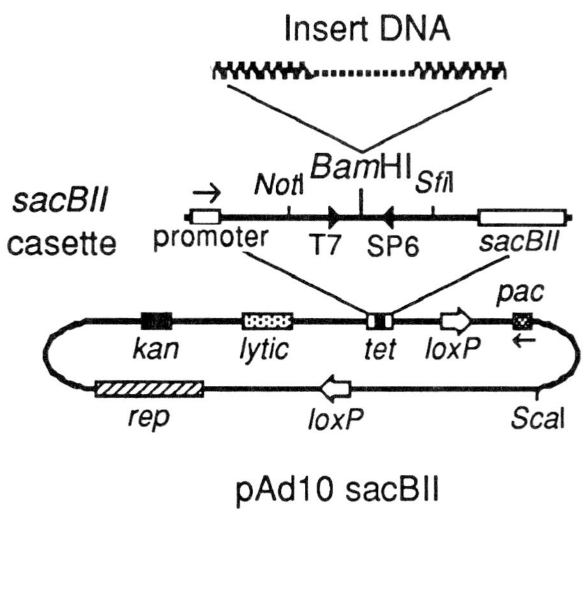

Fig. 2.2. Diagram of the P1 vector pAd10 sacBII. Data from Pierce JC, Sauer B, Sternberg N, Proc Natl Acad Sci USA 1992; 89:2056-2060.

The *tet* gene in pNS582-tet14 Ad10 (Fig. 2.1) contains two unique restriction sites (*Bam*HI and *Sal*I) suitable for use as cloning sites. Insertion of exogenous DNA into either site inactivates the *tet* gene and results in tetracycline sensitivity, which is used as a phenotypic attribute among kanamycin-resistant transfectants to identify those with insertions into *tet*. The pAd10 sacBII vector (Fig. 2) contains a *Bam*HI cloning site that provides positive selection for clones with inserts. The *Drosophila* libraries produced in these vectors were constructed using *Sau*3A genomic fragments obtained by partial digestion inserted into the *Bam*HI site of either pNS582-tet14 Ad10 or pAd10 sacBII. *Sau*3A cleaves at the slash in 5'-/GATC-3' and *Bam*HI cleaves at the slash in 5'-G/GATCC-3', producing complementary 5' overhangs; therefore, approximately 1/16 of the *Sau*3A-*Bam*HI vector-insert junctions will have a reconstituted *Bam*HI site.

In cloning with either pNS582-tet14 Ad10 or pAd10 sacBII, the vector is prepared by banding in a cesium chloride density gradient and cleaved with both *Sca*I and *Bam*HI. The *Sca*I cleavage opens the vector at the unique *Sca*I site clockwise from the *pac* site, and the *Bam*HI cleavage frees the short vector arm containing the *pac* site. The digestion with *Sca*I must be carried to virtual completion, otherwise a significant proportion of clones with inserts in the size range 30 kb to 40 kb may be recovered.[6]

After the vector has been digested and the terminal phosphate removed from the molecules by treatment with alkaline phosphatase, *Sau*3A fragments of genomic DNA are ligated onto the vector arms. Ligation products joining two *kan*-containing arms lack a *pac* site and so are not capable of packaging into phage heads in the subsequent steps. Ligation products joining two *pac*-containing arms lack the P1 rep and lytic replicons and other essential elements and so cannot be maintained stably even if they should happen to be packaged.

The desired products of the ligation each join a *kan*-containing vector arm and a *pac*-containing vector arm flanking a genomic *Sau*3A fragment. In the next step in the process, the ligated DNA is mixed with an extract of cells containing the pacase cleavage proteins, which cleave the linear molecule at the *pac* site. The resulting DNA is mixed with a second cell extract containing phage heads and tails but lacking the pacase proteins. This extract carries out headful packaging starting with the *pac* site of each linear molecule and proceeding until the head is filled and the molecule is cleaved approximately 110 kb counterclockwise (relative to the orientation in Figs. 2.1 and 2.2) from the *pac* site. In order that the headful of DNA will include the element essential for P1 replication (*rep*) and the downstream *loxP* site necessary for circularization, the insert of genomic DNA must be in the size range from 70 to 90 kb. After the headful cleavage has been made, the phage tails assemble spontaneously to create an infectious particle.

Following packaging, the infectious particles are use to transfect a strain of *E. coli* that expresses the Cre recombinase constitutively. Presence of the Cre recombinase catalyzes a site-specific recombination between the *loxP* sites on the linear DNA that is injected into the cells, with the result that the vector is circularized at the sites of *loxP* with elimination of sequences that were upstream and downstream from them (Fig. 2.3). The usual *E. coli* strains used for transfection also contain mutations in *recA*, which inactivates the major system of recombination and minimizes the risk of recombination between possibly repeated sequences in the insert DNA; *lacI*^{q}, which represses activation of the lac-driven P1

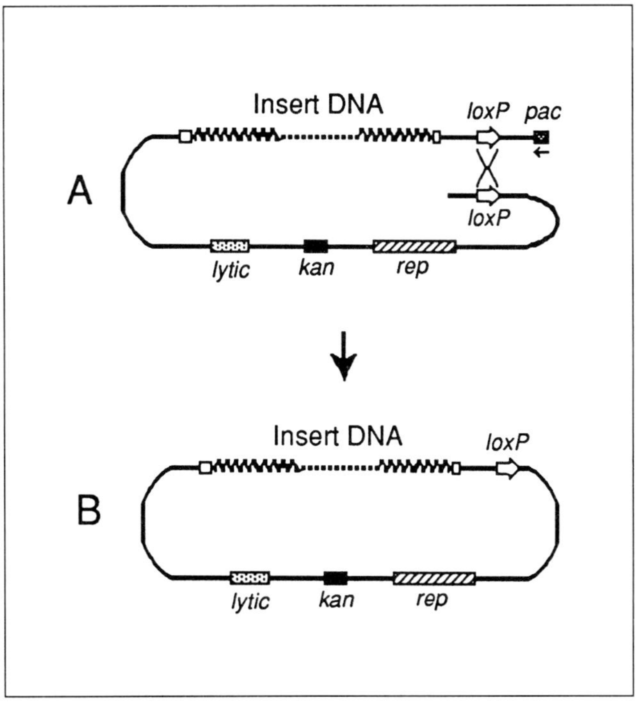

Fig. 2.3. Circularization of the P1 vector at the loxP sites.

lytic replicon; and *mcrAB*, which reduces loss of insert sequences containing methylated C (a feature not relevant to *Drosophila*, which has no detectable level of methylated C in its DNA). For the vector pNS582-tet14 Ad10, we used the host strain NS3145 [*recA hsdM*$^+$ *hsdR*$^-$ *mcrA mcrB* F'*lacI*^{q} (λ*imm*434 *nin*5 X1-cre)] and the transfected cells were plated on LB agar plates containing 25 μg/ml kanamycin.[9] For the vector pAd10 sacBII, we used the host strain NS3529 [*recA mcrA* Δ(*mcrB, mrr, hsdR, hsdM*) λ*imm* 434 *nin*5 X1-cre and λ*imm*λLP1], which carries the *lacI*^{q} and *cre* genes on separate integrated prophages;[6] the transfected cells were plated on LB agar plates containing 25 μg/ml kanamycin and 5% sucrose.[10]

The packaging efficiency of the P1 system can yield greater than 10^4 clones per μg of source DNA[11] as compared with a typical yield of 10^2-10^3 clones per μg of source DNA for the YAC cloning system.[12] Although molecules smaller than 75 kb can be packaged in P1 heads, these account for as few as 10% of the clones owing to specific modifications incorporated into the P1 vector.[11]

It should be noted here that a somewhat modified P1 system has also been developed that permits inserts in the size range 130-150 kb to be cloned.[13] The vector, called pCYPAC1, is a modified form of pAd10 sacBII in which the adenovirus fragment has been removed and in which a modified pUC19 plasmid has been inserted into the *Bam*HI cloning site. The insertion still permits positive selection of recombinant clones by means of *sac*BII selection, but it also reduces the toxicity of *sac*BII expression in the absence of sucrose selection.[13] Using pCYPAC1, the ligation products are introduced into *E. coli* by elecroporation rather than by packaging and bacteriophage infection.

PREPARATION OF VECTOR DNA

With the foregoing as the theoretical background, we now present additional detail regarding the procedures followed in constructing the *Drosophila* P1 libraries.[9,10,14] The methods apply to both the pNS582-tet14 Ad10 and the pAd10 sacBII libraries.

Vector DNA was prepared by banding in a cesium chloride density gradient and cleaved to completion with *Sca*I. To minimize concatemer formation, an aliquot containing 1 μg of vector cleaved with *Sca*I was treated with a tenfold excess of calf intestinal alkaline phosphatase in 40 µl of 50 mM Tris-HCl at pH 8.0 for 45 minutes, heated for 10 minutes at 68° C, extracted in phenol/chloroform, and precipitated in ethanol. The resuspended DNA was then digested to completion with *Bam*HI and treated with calf intestinal alkaline phosphatase in sufficient quantity to eliminate detectable religation of the vector arms in test aliquots. The solution containing the digested and phosphatased vector arms was heated for 10 minutes at 68° C, extracted in phenol/chloroform, precipitated in ethanol, and the DNA resuspended in 50 mM Tris-HCl at pH 8.0.

PREPARATION AND LIGATION OF GENOMIC DNA

Drosophila P1 libraries were prepared with genomic DNA from an isogenic strain of genotype *y* (yellow body); *cn* (cinnabar eyes) *bw* (brown eyes) *sp* (speck cuticular pigment) provided by J. Kennison. Adult flies totaling 0.5 grams were homogenized in NIB buffer (10 mM Tris, 10 mM NaCl, 10 mM EDTA 0.15 mM spermine, 0.15 mM spermidine, 0.5% triton X-100, pH 8.5). The nuclei were sedimented by centrifugation and the nuclear pellet washed twice with NIB and lysed in NIB with 3% sarkosyl. Following lysis, the DNA was banded in a CsCl gradient.[15] The viscous, high molecular weight band was isolated from the gradient and dialyzed three times against TE buffer (10 mM Tris-HCl, 1 mM EDTA, pH 8.0).

The large fragments of genomic DNA and partially digested with *Sau*3A. Optimal conditions were determined in 20 µl aliquots of DNA at 100 ng/µl mixed with 4 µl of 10 × restriction-digest buffer without magnesium (100 mM Tris-HCl, 1 M NaCl, 1 mg/ml bovine serum albumin), 13.5 µl of distilled H_2O, and 1 µl of *Sau*3A diluted in 1 × restriction-enzyme buffer to concentrations of 0.1, 0.001, 0.0001 U (Boehringer Mannheim). After equilibration at 4° C overnight, the three reactions, plus a control with no restriction enzyme, were incubated for 30 minutes at 37° C in the presence of 10 mM $MgCl_2$ and heated for 10 minutes at 68° C. The degree of digestion was analyzed using CHEF (contour-clamped homogeneous electric field) gel electrophoresis.[16] The enzyme concentration producing the largest molar quantity of DNA fragments in the size range 70-150 kb was repeated with a five-fold increase in quantity of genomic DNA, restriction buffer, distilled H_2O, and enzyme in order to provide sufficient quantities of partially digested genomic DNA for the ligations.

After *Sau*3A treatment, 400 µl of the partially digested DNA was layered onto 18 ml sucrose gradients (10-40% sucrose) and centrifuged in a AH627 rotor for 20 hr at 29,100 × g. Gradients were collected from the bottom of the centrifuge tube in 300 µl fractions and sample aliquots were analyzed by CHEF gel electrophoresis. Fractions containing DNA in the size range 75-100 kb were dialyzed for 1.5 hours at 25° C on filters (VSWP 02500, Millipore) floated on TE buffer. Fractions were concentrated with *sec*-butanol and either dialyzed again or precipitated in ethanol.

Ligations between genomic DNA fragments and vector DNA were carried out in an equal molar ratio of genomic DNA and vector DNA in the presence of bacteriophage T4 DNA ligase (Life Technologies, Inc.) in 1 × ligation buffer (0.05 M Tris-HCl, 0.01 M dithiothreitol, 1mM ATP, 0.01 $MgCl_2$).

PACKAGING, TRANSFECTION, AND RECOVERY OF CLONES

Pierce and Sternberg[6] give detailed instructions for preparation of the packaging extract and the packaging of ligated DNA. Following packaging, *E. coli* cells were infected with freshly packaged bacteriophage and plated. For the pNS582-tet14 Ad10 library, the host strain was NS3145 and the cells were plated on LB plates[17] plus 25 µg/ml kanamycin. Kanamycin-resistant colonies were tested for tetracycline resistance, and tetracycline-sensitive bacterial colonies were inoculated into 0.5 ml LB medium and grown overnight at 37° C. Glycerol was added to the cultures to a final concentration of 15-20%, and 100 µl of each culture was dispensed into a well of a 96-well microtiter dish for long term storage and the preparation of filters. For the pAd10 sacBII libraries, the host strain was NS3529 and the cells were plated on LB agar containing 25 µg/ml kanamycin and 5% sucrose. Surviving cells were picked and inoculated into 0.5 ml LB medium, grown overnight at 37° C, and stored in glycerol as described.

Table 2.1. Number of clones arrayed in P1 libraries

P1 vector	Type of insert	Number of clones
pNS582-tet14 Ad10	*Sau*3A partial	3,840
pAd10 sacBII	*Sau*3A partial	5,376

Number euchromatic genome equivalents approximately 5.0

ORGANIZATION OF THE P1 LIBRARIES

The initial library in pNS582-tet14 Ad10 consisted of 3,840 clones arrayed in the wells of 40 microtiter dishes.[9] A second library in pAd10 sacBII consisted of 5,376 clones arrayed in 56 microtiter dishes.[10] These 9,216 clones in pNS582-tet14 Ad10 and pAd10 sacBII constitute "the" *Drosophila* P1 library (Table 2.1). An additional set of 9,216 clones was also picked from the pAd10 sacBII library by a robot arm at the Lawrence Berkeley Laboratories; these additional clones serve as a reserve library for completing contigs and other purposes as needed.

The nomenclature of the clones warrants comment. The original clones were numbered according to their position in the wells of the microtiter plates. In each plate there are 8 horizontal rows and 12 vertical columns, and the clone in row i (i = 1 to 8) and column j (j = 1 to 12) is numbered k = (i-1)*8 + j. Thus, from left to right, the clone numbers of clones in the successive rows are 1-12, 13-24, 25-36, 37-48, 49-60, 61-72, 73-84, and 85-96. In the entire library, each clone is given a hyphenated name in which the first number is the plate number in the library and the second is the number of the clone in the plate; thus, clone 40-16 is clone number 16 in plate 40. We also prefix the clones in the original library with the letter C (for Cambridge), yielding the name C40-16 for the clone just mentioned.

A small proportion of the clones in the P1 libraries (approximately 1%) are duplicates picked from the same colony on the selective plates. The duplicates, though small in number, present a problem for the localization of clones combinatorially by PCR amplification from DNA pools containing all clones in each row and all clones in each column. For example, if the same clone is in location C26-18 (row 2, column 6) and also in C26-43 (row 4, column 7), then the pool of all clones in plate 26 will be positive (first-level PCR screen), but so will the pools for columns 2 and 4 and the pools for rows 6 and 7 (second-level PCR screen); thus, it is not clear from the second-level PCR screen whether the true positive clones are in the (row, column) combinations (2, 6), (2, 7), (4, 6), or (4, 7), and an additional PCR reaction is necessary to make the determination. To minimize this problem, Michael Palazzolo at the Lawrence Berkeley Laboratories purified single colonies from each clone in the Cambridge library and reorganized them so that the 96 clones from each plate in the Cambridge library would, in the new organization (called the Berkeley or B library) be distributed among 96 different plates. Insofar as it was possible, the clone Ci-j in the Cambridge library became clone Bj-i in the Berkeley library, where again the first number is the plate number and the second is the well number in the plate. Then, for convenience of record keeping in the database (and also to pay off the debt of nomenclature promised to David Smoller), the clones were numbered consecutively with five digits preceded by the designation DS. The DS numbers are the official names of the clones and the hyphenated "B" numbers are the locators for finding the clone in the Berkeley library. For any clone in the Berkeley library at location i-j (i = 1 through 96 and j = 1 though 96), the DS number is equal to (i-1)*96 + j. Clones C81-j (j = 1 to 96) and C131-j (j = 1 to 96), which are not present in the Berkeley library, have been assigned DS numbers DS09217 through DS09312 and DS09313 through DS09408, respectively, as if they were Berkeley plates 97 and 98.

All of the clones that had previously been localized by in situ hybridization had to be included in the restreaked and reorganized Berkeley library, and so there is not a perfect one-to-one correspondence between the Cambridge and Berkeley numbers. Table 2.2 provides a key between the Berkeley and the Cambridge nomenclatures. The clones in Flybase, which is accessible by anonymous FTP (file transfer protocol) from ftp.bio.indiana.edu in the directory flybase or by electronic mail requests to flybase@morgan.harvard.edu, are named according to their position in the Berkeley library (Bi-j) and their DS number.

SIZE DISTRIBUTION OF P1 INSERTS

In P1 clones that have been circularized at the *loxP* sites by the Cre recombinase, the total length of the remaining vector sequence is 14.6 kb for pNS582-tet14 Ad10 and 16.6 kb for pAd10 sacBII. These lengths of vector correspond to the aggregate of fragment lengths produced by *Bgl*II and *Xho*I double digests of the circularized vectors without inserts.[6] The pNS582-tet14 Ad10 vector yields *BglII—XhoI* double digest fragments of 0.9 kb, 1.9 kb, 3.8 kb, and 8.0 kb, totaling 14.6 kb (in P1 clones, the 8.0 kb fragment is interrupted by the insert). The pAd10 sacBII vector yields *BglII—XhoI* double digest fragments of 0.9 kb, 1.9 kb, 3.8 kb, and 10.0 kb, totaling 16.6 kb (in P1 clones the 10.0 kb fragment is interrupted by the insert).

Table 2.2. Nomenclature of P1 clones

Berkeley library*	Cambridge library
Plate i (for i = 1 to 96), clone j (for j = 1 to 80)	Clone Cj-i
Plate i (for i = 1 to 96), clone j (for j = 82 to 88)	Clone Cj-i
Plate i (for i = 1 to 96), clone j = 89	Clone C90-i
Plate i (for i = 1 to 96), clone j = 90	Clone C91-i
Plate i (for i = 1 to 96), clone j = 91	Clone C92-i
Plate i (for i = 1 to 96), clone j = 92	Clone C135-i
Plate i (for i = 1 to 96), clone j = 93	Clone C136-i
Plate i (for i = 1 to 96), clone j = 94	Clone C137-i
Plate i (for i = 1 to 96), clone j = 95	Clone C138-i
Plate i (for i = 1 to 96), clone j = 96	Clone C139-i

* The DS number for each clone is a five-digit number (left justified with zeros) derived as follows. For plate number *i*, clone *j* (*j* = 1 to 96 in dictionary order) in the Berkeley library, the DS number is (*i* - 1)*96 + *j*.

The size of inserts in P1 clones can be determined by digestion with restriction enzymes such as *Eco*RI, *Bam*HI, or *Bgl*II. Resulting fragments smaller then 20 kb are separated in conventional agarose gels,[18] and those larger than 20 kb in CHEF gels.[16] The vector pNS582-tet14 Ad10 contains no cleavage sites for *Not*I, and therefore the size of clones containing unique *Not*I sites within the insert can be determined by *Not*I digestion and CHEF gel electrophoresis. Assuming an average insert size of 80 kb, the expected proportion of *Drosophila* clones with unique *Not*I sites in the insert is expected to be 36%. In one small test we observed eight such clones among 16.[9] *Not*I and *Sfi*I can be used to determine the sizes of inserts in the vector pAd10 sacBII (Fig. 2.2) because this vector contains unique cleavage sites for *Not*I and *Sfi*I flanking the cloning site. These separations are carried out in 20 cm x 20 cm, 1% agarose gels in 0.5 TBE buffer (0.045 Tris-borate, 0.001 M EDTA, pH 8.3) in a BioRad CHEF-DR II apparatus for CHEF gel electrophoresis.[14] However, complete digestion with *Not*I is sometimes difficult.[6]

Figure 2.4 illustrates the size distribution of inserts of *Drosophila* DNA among 25 clones in pNS582-tet14 Ad10 and 20 clones in pAd10 sacBII. The average size of insert in the former is 83.0 kb and that in the latter is 82.5 kb. These values are close to the theoretical expectations. A small proportion of clones in the pAd10 sacBII library, estimated at fewer than 1%, appear to lack inserts of *Drosophila* DNA; another small proportion of clones contain inserts smaller than 60 kb.

P1 LIBRARY FROM *DROSOPHILA VIRILIS*

Our laboratory has been particularly interested in applications of genome analysis to evolutionary studies. To provide cloned material suitable for evolutionary studies in *Drosophila*, we also produced a P1 library from *D. virilis*. This species is in the subgenus *Drosophila* (*D. melanogaster* is in the subgenus *Sophophora*) and has been the object of substantial genetic and evolutionary study.[19] Details of the construction of the *D. virilis* P1 are described in Lozovskaya et al.[14] The source of genomic DNA was *D. virilis* strain 9, a wildtype strain collected in 1970 in Batumi, Georgia (former USSR), and maintained in laboratory culture since that time. The hybridizations in situ were also carried out with larvae from strain 9. High-molecular weight DNA was prepared as described above and cloned into the P1 vector pAd10sacBII. After transfection of the host strain NS3529 and plating on LB agar containing 25 μg/ml kanamycin and 5% sucrose, the surviving colonies were individually arrayed in the wells of 105 microtiter plates for a total of 10,080 clones.

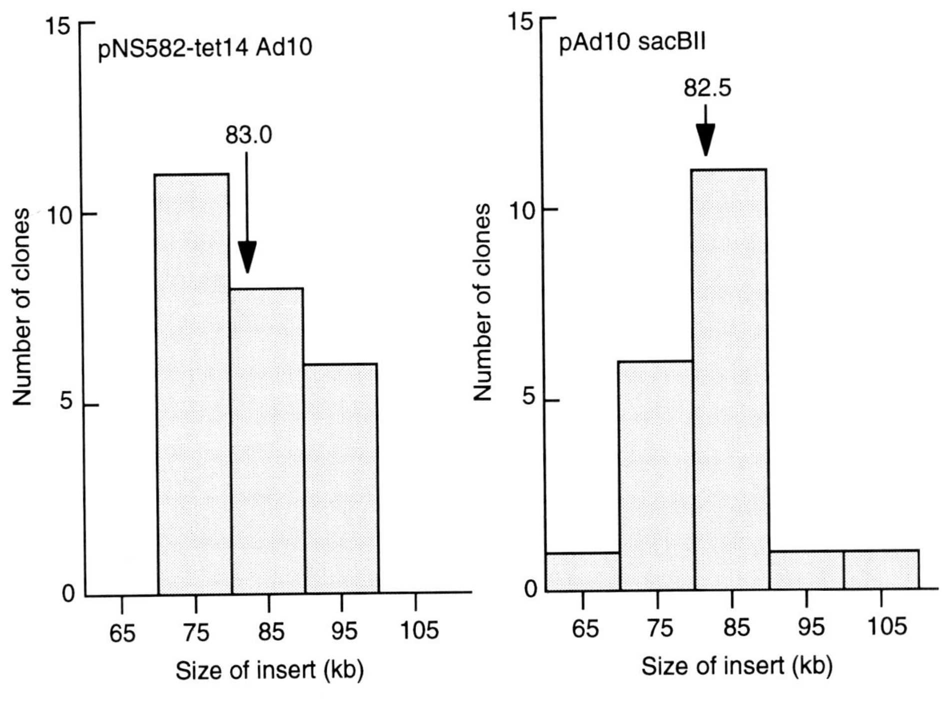

Fig. 2.4. The size distribution of inserts among clones of DNA from D. melanogaster *in pNS582-tet14 Ad10 and pAd10 sacBII.*

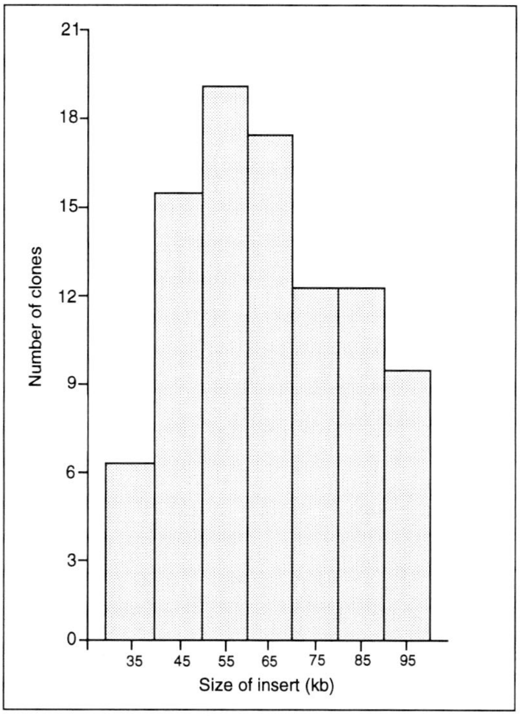

Fig. 2.5. The distribution of insert sizes among P1 clones in pAd10 sacBII from D. virilis.

The size of the insert in each of 90 randomly selected clones was estimated from the fragment sizes produced by digestion with *Eco*RI, *Bam*HI, or *Bgl*II. A subset of clones was checked independently by digesting with either *Not*I or *Sfi*I and separating the fragments in CHEF gels. The distribution of insert sizes among the *D. virilis* P1 clones is given in Figure 2.5. The average insert size is 65.8 kb, which is somewhat smaller than in the *D. melanogaster* P1 clones. On the other hand, approximately 75% of the clones have inserts exceeding 50 kb, and approximately 25% have inserts exceeding 80 kb.

YAC VECTORS

At the very beginning of the *Drosophila* genome project, Dan Garza had created a special YAC vector in which the cloning site was flanked by the ends of the transposable element *P* to allow for the possibility of germline transformation with large DNA fragments. This vector (called pYACP-1) was derived from the first-generation vector pYAC2, which is shown in Figure 2.6.[20] The pYAC2 vector contains the CEN4 centromeric sequence from yeast (*Saccharomyces cerevisiae*) chromosome 4 and telomeric sequences from *Tetrahymena.* The vector also contains an *E. coli* origin of replication (*ori*) and an ampicillin-resistance gene (*amp*). For replication in yeast, the vector contains an autonomously replicating sequence (ARS1) and the selectable markers *TRP1* (tryptophan) and *URA3* (uridine). The *Sma*I cloning site in pYAC2 is present in the coding region of the nonsense suppressor *SUP4,* and insertion of a large DNA fragment in the cloning site inactivates the suppressor and allows detection of the insert-containing clones. The *Bam*HI fragment between the telomeres carries a yeast *HIS3* gene, but this fragment is cleaved and discarded in the production of YACs. The approximate positions of the unique *Cla*I and *Sal*I cleavage sites in pYAC2 are also shown because these sites were used in the construction of pYACP-1.

The vector pYACP-1 is diagrammed in Figure 2.7 and its derivation is described in Garza et al.[12] In addition to having the functional elements necessary for the selection and maintenance of clones in yeast, pYACP-1 contains the ends of the *Drosophila* transposable element *P* and the *hsp70:G418*-resistance gene. The 5' end of the *P* element was isolated as a 1.3 kb *Eco*RI—*Sal*I fragment from the vector ep19.1;[21] this fragment also contains sequences from the *white* and *rosy* genes upstream of the 5' end of *P*. The 3' end of the *P* element was isolated as a 1.9 kb fragment from an *Xmn*I

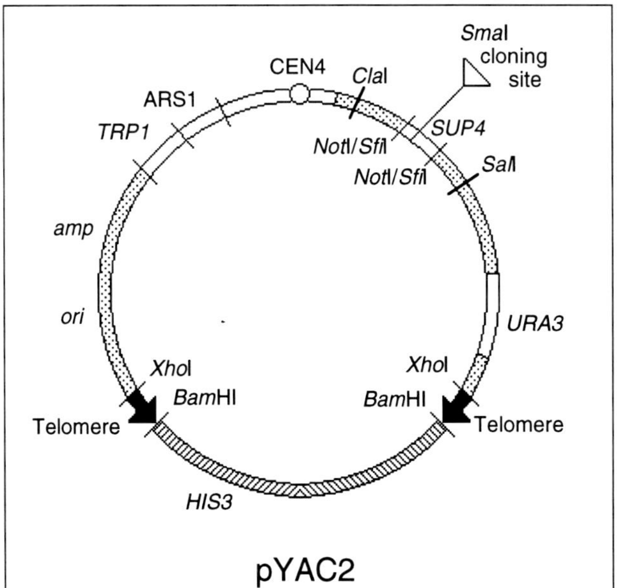

Fig. 2.6. Diagram of the YAC cloning vector pYAC2. Data from Burke DT, Carle GF, Olson MV. Science 1987; 236:806-812.

(complete digestion)—*Bal*I (partial digestion) treatment of the transformation vector pUChsneo;[22] this fragment also contains the *hsp70:G418*-resistance gene downsteam from the 3' end of *P*. After manipulation of these fragments so as to flank the 1.3 kb fragment with *Sal*I sites and the 1.9 kb fragment with *Cla*I sites,[12] the 1.3 *Sal*I fragment was ligated into the *Sal*I site of pYAC2; the resulting vector was digested with *Cla*I and the 1.9 kb *Cla*I fragment inserted. Clones with the proper orientation of the *P*-element ends were identified with suitable restriction digests.

Other, more conventional YAC vectors were also used in the construction of the *Drosophila* libraries; these are derivatives of pYAC2 diagrammed in Figure 2.8. The pYAC3 vector is one in which oligonucleotide site-directed mutagenesis was used to create a unique *Sna*BI site within the small intron of *SUP4* and to eliminate the *Not*I and *Sfi*I sites flanking the *SUP4* gene.[20] The pYAC4 (*Eco*RI) and pYAC5 (*Not*I) vectors were derived from pYAC3 by insertion of an *Eco*RI linker or a *Not*I linker, respectively, into the *Sna*BI site.[20]

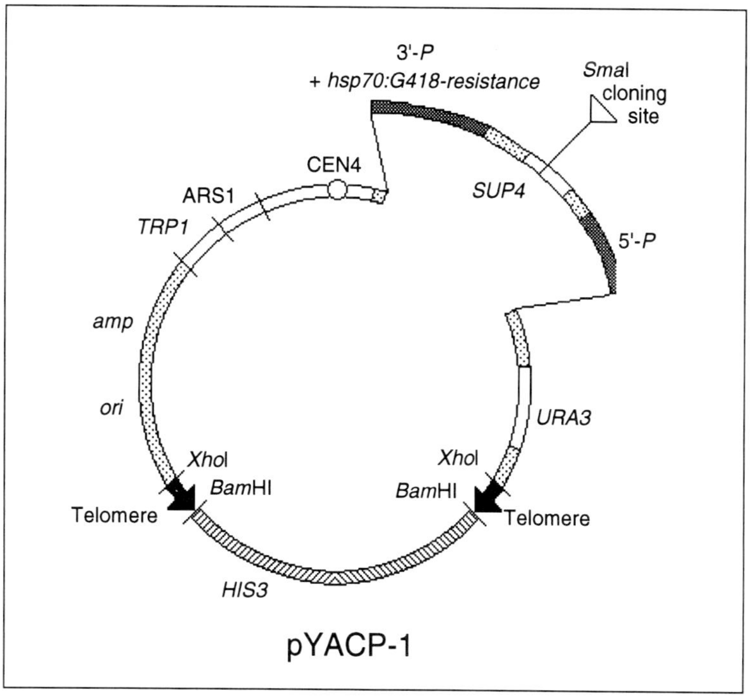

Fig. 2.7. Diagram of the YAC cloning vector pYACP-1. Data from Garza D, Ajioka JW, Burke DT, Hartl DL.,Science 1989; 246:641-646.

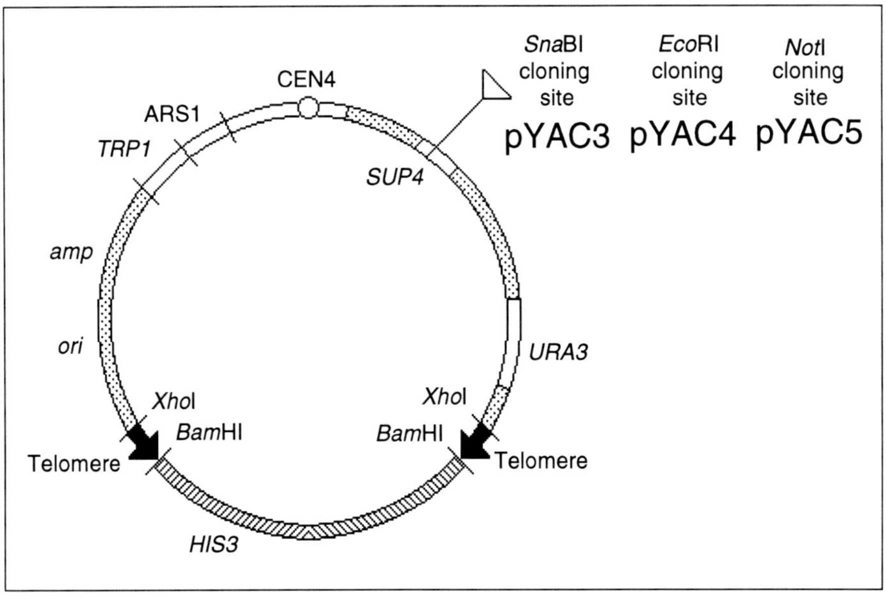

Fig. 2.8. The YAC vectors pYAC3, pYAC4 (cloning site EcoRI*), and pYAC5 (cloning site* NotI*). Data from Burke DT, Carle GF, Olson MV, Science 1987; 236:806-812.*

PRINCIPLES OF YAC CLONING

The selection of YAC clones of the desired structure is based on the presence of genetic markers in each arm of the YAC and the inactivation of the *SUP4* suppressor by insertion of exogenous DNA. YAC vector DNA was prepared by digestion with *Bam*HI and the restriction enzyme that cleaves at the cloning site (*Sma*I, *Eco*RI, and *Not*I for pYACP-1, pYAC4, and pYAC5, respectively).[20] The resulting three fragments were treated with alkaline phosphatase to prevent religation. After addition of exogenous DNA and ligation, the desired products are those containing the *TRP1*-CEN4 fragment (forming the left arm of the YAC) and the *URA3* fragment (forming the terminus of the right arm of the YAC) flanking a large fragment of exogenous DNA; in theory, these are the only products that are genetically stable.

YAC clones were obtained by transformation of yeast spheroplasts that were then embedded in selective medium. The host strain was AB1380,[20] which carries *MAT*α ψ^+ *ura3 trp1 ade2-1 can1-100 lys2-1 his5*. The primary selection was for growth in the absence of uridine and tryptophan on medium adjusted to pH 5.8 containing, per liter, 182 gm sorbitol, 1.7 gm yeast nitrogen base, 5.0 gm $(NH_4)_2SO_4$, 20 gm glucose, 18 gm bacto-agar, and the following: 10 mg adenine hemisulfate, 40 mg arginine-HCl, 20 mg histidine-HCl, 60 mg isoleucine, 60 mg leucine, 50 mg lysine-HCl, 20 mg methionine, 50 mg phenylalanine, 200 mg threonine, and 50 mg tyrosine.[23] Growth in the absence of uracil selects for *URA3*, which marks the right arm of the YAC and complements the *ura3* mutation in the host; growth in the absence of tryptophan selects for *TRP1*, which marks the left arm of the YAC. The *SUP4* suppressor is an ochre-suppressing $tRNA^{Tyr}$ that allows a convenient colony-color test of its presence. Yeast cells containing an *ade2-ochre* mutation (*ade2-1* in the strain AB1380) are red in the absence of suppression and white in the presence of suppression, hence insertional inactivation of *SUP4* results in red colonies. Presence of the cytoplasmic factor ψ^{+} in the host enhances the suppression and also the red/white color difference between the colony types. The desired recombinants are also resistant to canavanine by virtue of the *can1-100* marker in the host.

CONSTRUCTION OF *DROSOPHILA* YAC LIBRARIES

The YAC clones used for in situ hybridization were derived from different YAC libraries in which the *Drosophila* DNA fragments were prepared in different ways.

The clones prefixed "DY" are contained in the vector pYACP-1 (Fig. 2.7). They were derived from randomly sheared DNA from the *Drosophila* stock Oregon RC.[12] DNA was isolated from lysed nuclei and purified by CsCl density gradient centrifugation, which yielded DNA molecules up to approximately 500 kb in length. The randomly sheared DNA ends were repaired with bacteriophage T4 DNA polymerase and fragments were separated by size in sucrose velocity gradients.[24] Fractions were collected and assayed by field inversion gel electrophoresis, and fractions with DNA fragments above 120 kb were pooled and concentrated in a collodion bag. Size-fractionated DNA was used as insert DNA for ligation with prepared pYACP-1 vector arms, size-fractionated and concentrated again, and then used to transform spheroplasts of yeast strain AB1380.

The clones designated "N" contain *Not*I fragments cloned into pYAC5 (Fig. 2.8) by Andrew Link according to the method described in Danilevskaya et al.[25] In these libraries, DNA was obtained from gastrula cells of Oregon RC. The cells were washed in Hank's balanced salt solution at 4° C, centrifuged, and resuspended to a concentration of 9×10^8 cells/ml. Dilutions of 0.05–0.1 ml were warmed to 25° C, mixed with 1% low melting temperature agarose in 1 M NaCl, 10 mM Tris (pH 7.5) in a ratio of cell suspension/agarose of 4:5 (v/v), and aliquots were transferred to acrylic molds to form agarose plugs. The plugs were pushed out of the mold into a lysis solution of 1% Sarkosyl, 0.5 M EDTA, 2 mg/ml proteinase K, 1 mM Tris (pH 8), then sealed in plastic bags and incubated at 50° C for 48 hours with gentle shaking. The plugs were washed three times for one hour each at 4° C in a buffer of 1 mM EDTA, 10 mM Tris (pH 8), followed by three washes in *Not*I restriction buffer.

The *Not*I digestion was carried out in 600 µl of restriction buffer with gentle agitation at 37° C for 8-12 hours. The plugs were transferred to 1.5 ml tubes, vector arms added in an equal mass as genomic DNA, and incubated at 70° C for 5-10 minutes to melt the agarose. An equal volume of ligation buffer at 4° C and 50 Weiss units of T4 DNA ligase was added to each tube and incubations were carried out for 12-24 hours at 14° C prior to transformation.

The YAC clones prefixed "E," "R," or "RX" are in the vector pYAC4 (Fig. 2.8) and contain *Eco*RI fragments resulting from partial *Eco*RI digestion of DNA from a Canton S stock. The "E" clones were prepared by Elena Lozovskaya and the "R" and "RX" clones by Heini Cai.[26] The "E" clones were prepared from genomic DNA fragments as described above for the pYACP-1 library, except that the high molecular weight DNA was lightly digested with *Eco*RI prior to size fractionation to enrich for the size range 150–550 kb. After size fractionation in sucrose velocity gradients, the *Eco*RI fragments were ligated onto the vector arms of pYAC4 and transformation carried out with individual fractions of increasing fragment size. The "R" and "RX" clones were prepared as described by Cai et al.[26] Clones designated "Rt" also contain *Eco*RI fragments in pYAC4 and were derived from a stock of genotype T(Y;2)CB25, *cn; y, cn bw* provided by Terry Lyttle.[26]

A small number of clones, designated with the prefix "T," were derived from *Eco*RI partial digests of Oregon RC DNA using the "half-YAC" vector designed to select for telomere addition sites in yeast.[27] This vector yields ligation products lacking the right arm of the YAC, and so any DNA fragment attached to the left arm that can serve as a telomere-addition site is able to rescue the YAC and confer genetic stability. The *Drosophila* DNA fragments that could rescue the "T" clones proved not to derive from the tips of the chromosomes, however; upon in situ hybridzation, most of the clones were localized in the underreplicated mass of heterochromatin that aggregates to form the chromocenter or, less frequently, in euchromatic sites in the chromosome arms. The failure of the half-YAC selection to yield telomeric clones is consistent with data from Levis et al[28] suggesting that *D. melanogaster* may have unconventional, transposon-related telomeres as compared with the simple tandemly repeated sequences found in most other eukaryotes.

SIZE OF INSERTS IN YAC CLONES

The *Drosophila* YAC libraries never achieved the large insert sizes present in some human libraries.[29,30] The "E" (*Eco*RI) library provides a typical example. The sizes of the clones were estimated by field inversion gel electrophoresis.[20,24] About 60% of the clones

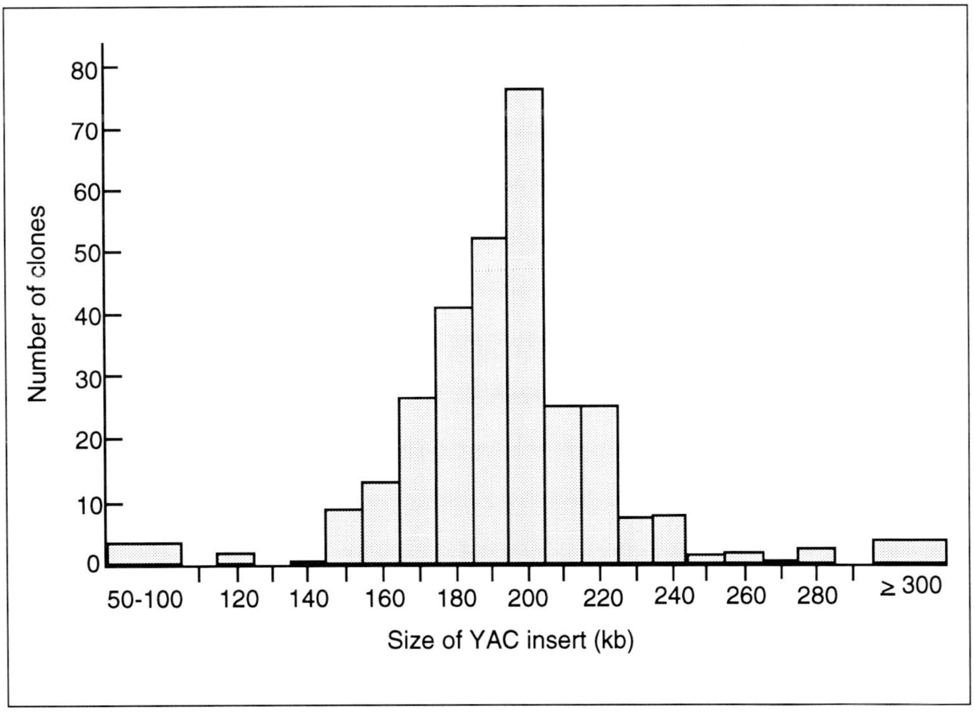

Fig. 2.9. The size distribution among 325 YAC clones of the "E" series in pYAC4 containing DNA fragments from D. melanogaster.

yielded a single, sharp YAC band after electrophoresis. The remaining clones contained YACs smaller than 50 kb, or contained two YACs, or produced a YAC band too faint to see and size reliably in the field inversion gel. The size distribution observed among 325 "E" clones with single, large DNA inserts is shown in Figure 2.9. The modal size and the mean size both equal 200 kb. Only 2.5% of the clones have inserts larger than 300 kb, and the maximum insert size in this sample is 470 kb.

COMPOSITION OF YAC LIBRARIES

From each of the YAC libraries, clones containing single, large DNA inserts were selected for incorporation into the permanent collection. The composition of clones according to vector in the "the" *Drosophila* YAC library is shown in Table 2.3. Assuming an average insert size of 200 kb, the 7,488 clones in Table 2.3 contain an aggregate of approximately 1.5×10^9 base pairs of DNA—the equivalent of about 9.1 haploid genomes of *Drosophila* DNA. About 10% of the clones are in the vector pYACP-1 (randomly sheared fragments), 36% in pYAC5 (*Not*I fragments), and 54% in pYAC4 (*Eco*RI fragments). Clones containing the largest inserts were chosen for localization by in situ hybridization.

Table 2.3. Number of clones arrayed in YAC libraries

YAC vector	Type of insert	Number of clones
pYACP-1	Randomly sheared	768
pYAC5	*Not*I partial	2,688
pYAC4	*Eco*RI partial	4,032

REFERENCES

1. Sidén-Kiamos I, Saunders RDC, Spanos L, Majerus T, Treanear J, Savakis C, Louis C, Glover DM, Ashburner M, Kafatos FC. Towards a physical map of the *Drosophila melanogaster* genome: Mapping of cosmid clones within defined genomic divisions. Nucleic Acids Res 1990; 18:6261-6270.

2. Kafatos FC, Louis C, Savakis C, Glover DM, Ashburner M, Link AJ, Sidén-Kiamos I, Saunders RDC. Integrated maps of the *Drosophila* genome: progress and prospects. Trends Genet 1991; 7:155-161.
3. Lehrach H, Drmanac R, Hoheisel J, Larin Z, Lennon G, Monaco AP, Nizetic D, Zehetner G, Poustka A. Hybridization fingerprinting in genome mapping and sequencing. In: K. E. Davies and S. M. Tilghman, eds. *Genome Analysis Volume 1: Genetic and Physical Mapping*. Cold Spring Harbor, NY: Cold Spring Harbor Laboratory Press 1990; 39-81.
4. Hoheisel JD, Lennon GG, Zehetner G, Lehrach H. Use of high coverage reference libraries of *Drosophila melanogaster* for relational data analysis; a step towards mapping and sequencing the genome. J Mol Biol 1991; 220:903-914.
5. Hoheisel JD. Application of hybridization techniques to genome mapping and sequencing. Trends Genet 1994; 10:79-83.
6. Pierce JC, Sternberg NL. Using the bacteriophage P1 system to clone high molecular weight genomic DNA. Methods Enzymol 1992; 216:549-574.
7. Hoess RH, Abremski K. Interaction of the bacteriophage P1 recombinase Cre with the recombining site *loxP*. Proc Natl Acad Sci USA 1984; 81:1026-1029.
8. Pierce JC, Sauer B, Sternberg N. A positive selection vector for cloning high molecular weight DNA by the bacteriophage P1 system: Improved cloning efficacy. Proc Natl Acad Sci USA 1992; 89:2056-2060.
9. Smoller DA, Petrov D, Hartl DL. Characterization of bacteriophage P1 library containing inserts of *Drosophila* DNA of 75-100 kilobase pairs. Chromosoma 1991; 100:487-494.
10. Hartl DL, Nurminsky DI, Jones RW, Lozovskaya ER. Genome structure and evolution in *Drosophila:* Applications of the framework P1 map. Proc Natl Acad Sci USA 1994; 91:6824-6829.
11. Sternberg N. Bacteriophage P1 cloning system for the isolation, amplification, and recovery of DNA fragments as large as 100 kilobase pairs. Proc Natl Acad Sci USA 1990; 87:103-107.
12. Garza D, Ajioka JW, Burke DT, Hartl DL. Mapping the *Drosophila* genome with yeast artificial chromosomes. Science 1989; 246:641-646.
13. Ioannou PA, Amemiya CT, Garnes J, Kroisel PM, Shizuya H, Chen C, Batzer MA, de Jong PJ. A new bacteriophage P1-derived vector for the propagation of large human DNA fragments. Nature Genet 1994; 6:84-89.
14. Lozovskaya ER, Petrov DA, Hartl DL. A combined molecular and cytogenetic approach to genome evolution in *Drosophila* using large-fragment DNA cloning. Chromosoma 1993; 102:253-266.

15. Bingham PM, Levis R, Rubin GM. Cloning of DNA sequences from the white locus of *Drosophila melanogaster* by a novel and general method. Cell 1981; 25:693-704.
16. Vollrath D, Davis RW. Resolution of DNA molecules greater than 5 megabases by contour-clamped homogeneous electric fields. Nucleic Acids Res 1987; 15:7865-7876.
17. Miller JH. *Experiments in Molecular Genetics*. Cold Spring Harbor, New York: Cold Spring Harbor Laboratory 1972.
18. Sambrook J, Fritsch EF, Maniatis T. *Molecular Cloning: A Laboratory Manual*. Cold Spring Harbor, New York: Cold Spring Harbor Laboratory 1989.
19. Alexander ML. The genetics of *Drosophila virilis*. In: M. Ashburner and E. Novitski, eds. *The Genetics and Biology of Drosophila*, Volume 1c. New York: Academic Press 1976; 1365-1427.
20. Burke DT, Carle GF, Olson MV. Cloning of large segments of exogenous DNA into yeast by means of artificial chromosome vectors. Science 1987; 236:806-812.
21. Lis JT, Simon JA, Sutton CA. New heat shock puffs and beta-galactosidase activity resulting from transformation of *Drosophila* with an *hsp70-lacZ* hybrid gene. Cell 1983; 35:403-410.
22. Steller H, Pirrotta V. A transposable *P* vector that confers selectable G418 resistance to *Drosophila* larvae. EMBO J 1985; 4:167-171.
23. Brownstein BH, Nelson DL eds. *YAC Libraries: A User's Guide*. New York: W. H. Freeman 1994.
24. Carle GF, Olson MV. Separation of chromosomal DNA molecules from yeast by orthogonal-field-alternation gel electrophoresis. Nucleic Acids Res 1984; 12:5647-5663.
25. Danilevskaya ON, Kurenova EV, Pavlova MN, Link AJ, Koga A, Vellek A, Hartl DL. He-T family DNA sequences in the Y chromosome of *Drosophila melanogaster* share homology with the X-linked Stellate genes. Chromosoma 1991; 100:118-124.
26. Cai H, Kiefel P, Yee J, Duncan I. A yeast artificial chromosome clone map of the *Drosophila* genome. Genetics 1994; 136: 1385-1401.
27. Riethman HC, Moyzis RK, Meyne J, Burke DT, Olson MV. Cloning human telomeric DNA fragments into *Saccharomyces cerevisiae* using a yeast-artificial-chromosome vector. Proc Natl Acad SciUSA 1989; 86:6240-6244.
28. Levis RW, Ganesan R, Houtchens K, Tolar LA, Sheen F. Transposons in place of telomeric repeats at a *Drosophila* telomere. Cell 1993; 75:1083-1093.

29. Chumakov I, Rigault P, Guillou S, Ougen P, Billaut A, Guasconi G, Gervy P, LeGall I, Soularue P, Grinas L, Bougueleret L, Bellanné-Chantelot Lacroix B, Barillot E, Gesnouin P, Pook S, Vaysseix G, Frelat G, Schmitz A, Sambucy J-L, Bosch A, Estivill X, Weissenbach J, Vignal A, Riethman H, Cox D, Patterson D, Gardiner K, Hattori M, Sakaki Y, Ichikawa H, Ohki M, Le Paslier D, Heilig R, Antonarakis S, Cohen D. Continuum of overlapping clones spanning the entire human chromosome 21. Nature 1992; 359:380-359.
30. Cohen D, Chumakov I, Weissenbach J. A first-generation physical map of the human genome. Nature 1993; 366:698-701.

CHAPTER 3

The Drosophila Genome

Over 80 years of intensive research starting with Thomas Hunt Morgan in 1910 have provided an unparalleled amount of information on structure and function of the *Drosophila* genome. From a genetic point of view, *Drosophila* is among the most thoroughly characterized multicellular eukaryotes.

This chapter provides an overview of those aspects of the organization of the *Drosophila* genome that are relevant to genome analysis. Readers interested in other important features of the *Drosophila* genome—gene structure, function, mutation, evolution, genetic analysis, and so forth—should consult the review articles by Rubin[1] and Merriam et al.[2] More detail can be found in the summary book by Ashburner[3] and still more detail in the classic 12-volume compendium of *Drosophila* genetics edited by Ashburner and others (Academic Press). The current status of the field of *Drosophila* developmental genetics is covered in Bate and Martinez Arias.[4]

GENOME SIZE

The genome of *Drosophila melanogaster* is one of the smallest among multicellular eukaryotes. It is estimated to be approximately 165 megabases (Mb),[5,6] or about one-twentieth the size of human genome. The *Drosophila* genome is composed of only four pairs of chromosomes—the heteromorphic X and Y chromosomes as well as two pairs of large autosomes and one pair of small autosomes. The largest *Drosophila* chromosome (chromosome 3) is approximately the same size as the smallest human chromosome.

EUCHROMATIN AND HETEROCHROMATIN

About two-thirds of the haploid genome of *Drosophila* (110 Mb) is composed of euchromatin, which undergoes decondensation and becomes invisible to light microscopy during interphase. Most mutations are located in this portion of the genome. Euchromatin consists largely (about 80%) of single-copy DNA and includes the vast majority of coding sequences. The remaining approximately 20% of euchromatin consists of moderately repeated sequences that are mainly representative of different families of transposable elements.

CODING CAPACITY OF THE EUCHROMATIN

The issue of the protein-coding capacity of the *Drosophila* genome has been extensively discussed in the literature (see, for instance, John and Miklos[7]). How many genes are necessary for the development and function of a fly? What fraction of genomic DNA codes for these genes? One traditional approach to these issues combined an estimate of the average rate of mutation of particular genes with the observed genome-wide rate of lethal mutation and a guess as to the fraction of all genes that could mutate to a recessive lethal. The answer came out around 5,000 genes. A second traditional approach was based on counts of the number of bands in the polytene chromosomes in salivary glands, which also resulted in an estimate of about 5,000 genes based on the evidence then available supporting the supposition of one genetic complementation unit per chromosome band.[8,9] The general agreement in the estimates of gene number obtained from these approaches was taken as too striking to be coincidental. Furthermore, taking the estimates at face value, both approaches also suggested that much of the euchromatin in the *Drosophila* genome might be noncoding.

The much later discovery of introns as well as large 5' and 3' regulatory regions in some genes implied that genes can be significantly larger than the amount of DNA required solely for coding. The actual number of coding sequences in *Drosophila* can only be determined from complete genomic sequencing. In *Drosophila*, progress in genomic sequencing is not yet far enough along to allow an estimate of total gene number by extrapolation. (*Drosophila* genomic sequencing is the subject of chapter 7.) However, in the nematode *Caenorhabditis elegans*, analysis of 2.2 megabases of continuous genomic sequence in a relatively gene-rich region has revealed a coding density of one coding sequence per every 5 kb.[10] The nematode genome is 100 Mb, suggesting a maximum coding capacity of about 20,000 coding sequences. Given that the *Drosophila* euchromatic genome is also 100 Mb, a maximum coding capacity of 20,000 coding sequences may also apply.

The estimate from *C. elegans* suggests that the earlier opinion of the number of genes in *Drosophila* may be substantially too small. Other types of molecular data from *Drosophila* suggest that the traditional estimates may be off by a factor of perhaps three or four. For example, it has been concluded from reassociation kinetics that about 16,000 different transcripts are expressed during the life cycle of *Drosophila*.[11] If each different transcript is equated with a different coding function, then there would be about 16,000 genes. Looked at from a different angle, the average transcript corresponds to about 7 kb of euchromatin (compared with one coding sequence per 5 kb in *C. elegans*).

An estimate of approximately 7 kb per *Drosophila* coding unit has also been obtained using combined genetic and molecular approaches in the analysis of the region containing the genes *rosy* and *Ace*.[12-14] This 315 kb region gives rise to 43 transcripts—an average of one transcript every 7 kb. Furthermore, the average size of a primary transcript in *Drosophila* is about 5 kb (data, obtained from 40 GenBank entries), which includes the protein-coding exons as well as the introns and other nontranslated sequences necessary for correct RNA processing and function of processed mRNA. An average of perhaps 1 kb or so of the surrounding DNA landscape might also be essential for the regulation of gene expression. Thus, if *Drosophila* has 15,000 to 20,000 genes with an average DNA content per gene of 6 kb (including regulatory regions), then the amount of euchromatic DNA allocated to genetic functions (transcription and its regulation) is 90 Mb to 120 Mb. There are also other DNA sequences in *Drosophila* chromosomes that are not transcribed but that play important roles in such processes as recombination and dosage compensation.[15] Hence, not much euchromatic DNA is left for nongenetic functions—certainly not the supposed great excess of genetic material that troubled molecular biologists 20 years ago.

Beyond the coding problem, many other mysteries remain. Some species of *Drosophila* have a genome size considerably greater than that of *D. melanogaster.* For example, the euchromatic genome of *D. virilis,* discussed in chapter 8, is about 150 Mb as compared with about 110 Mb in *D. melanogaster.* How is the extra 40 Mb in *D. virilis* distributed? It seems unlikely that *D. virilis* has 36% more genes than *D. melanogaster.* Are the introns or the 5' and 3' regulatory regions of each gene uniformly longer than in *D. melanogaster*? Are there many more pseudogenes in *D. virilis*? Is the number of families of transposable elements and their average copy number so much greater in *D. virilis*? More generally, what are the mechanisms by which genome size grows, shrinks, or is maintained? What features of gene arrangement, gene size, and relative spacing are conserved in evolution? What is the size distribution of synteny groups as a function of time? These and other important issues of genome size and evolution were once matters only for speculation but are finally accessible to experimental investigation through physical mapping and genomic sequencing.

HETEROCHROMATIN

In addition to 110 Mb of euchromatin, the genome of *D. melanogaster* also contains approximately 50 kb of heterochromatin.[16] The term "heterochromatin" is a venerable word first introduced by Heitz[17] in 1928 to distinguish densely staining regions of chromosomes (pycnotic regions) from regions not so heavily stained. Constitutive heterochromatin refers to the part of the genome that remains relatively condensed during the cell cycle, replicates later in the S period than does euchromatin,[18] undergoes little or no meiotic recombination relative to its physical length[19,20] and, in general, contains multiple families of repetitive sequences with various levels of reiteration.[21,22] Heitz's cytological definition distinguishing euchromatin from heterochromatin has held up very well through the decades. Indeed, later investigations have revealed deep structural and functional differences between these two major categories of chromatin organization.

Heterochromatin in *Drosophila* is classified as either alpha heterochromatin, which surrounds the centromeres and consists largely of blocks of highly repetitive, simple-sequence, "satellite" DNA; or beta heterochromatin, which includes moderately repeated DNA sequences and forms a diffuse, indistinctly banded boundary with euchromatin in the arms of the polytene chromosomes in salivary glands.[22] Although there is a low density of genes in heterochromatin, relative to its physical length,[23] many genetic functions other than housing the centromere (kinetochore) have been assigned to the heterochromatic regions of all chromosomes. For example, the heterochromatin of the X and Y chromosomes contain blocks of tandemly repeated genes coding for ribosomal RNA (chapter 4). The Y chromosome, which is entirely heterochromatic, contains six genes essential for fertility in males.[24-27] The heterochromatin of chromosomes 2 and 3 contain about 30 genes necessary for viability.[28-31] Heterochromatic regions also play important roles in the structural organization of the interphase nucleus and the segregation of achiasmate chromosomes in meiosis I.

Heterochromatin is also associated with a phenomenon called position effect variegation (reviewed in Baker[32] and Spofford[33]). Genes normally present in euchromatin do not function well when brought into the vicinity of heterochromatin. Their pattern of expression differs from one cell to the next; if variation in gene expression is detectable phenotypically (as, for example, with certain genes affecting eye color), then the resulting phenotype is mottled or variegated—hence the term variegated (V-type) position effect. The variable expression of euchromatic genes relocated in or near heterochromatin is correlated with structural changes in the chromatin due to the presence of a special heterochromatic nonhistone protein HP1,[34,35] which is highly conserved in evolution.[36] HP1 is evidently just one of a set of chromosomal proteins specialized for heterochromatin. Because variegated position effects are observed when euchromatic genes are brought into or near heterochromatin, it might also be expected that heterochromatic genes brought into euchromatin would show some irregularity in expression. This is indeed the case. Heterochromatin does provide an essential chromatin environment for the proper regulation of expression of heterochromatic genes. When genes that normally reside in heterochromatin are relocated into a euchromatic position, their expression is severely affected.[37]

Several recent experimental results have led to a resurgence of interest in heterochromatin among researchers in the *Drosophila* community. First, the use of special stocks containing suppressors of position-effect variegation have made it possible to detect insertion of the transposable element *P* into heterochromatin.[38] This approach makes the heterochromatin amenable to genetic analysis using the powerful methods of transposon tagging and enhancer trapping, including such difficult genes as the Y chromosome fertility factors. Secondly, methods for physical analysis of the 1.3 Mb minichromosome Dp(1;f)1187 (including localization of the centromere) have been developed.[39] Third, studies of transposable elements have indicated that at least some families of transposable elements are present in heterochromatin in large blocks that are conserved among strains.[40] These results have all led to a similar conclusion from different angles: The heterochromatin is not a boring genetic wasteland of countless megabases of tandemly repeated satellite DNAs interrupted here and there by an occasional functional gene and an occasional defective transposable element in the heterochromatic "graveyard." Rather, the heterochromatin has a complex and interesting genetic structure consisting of "islands" of unique sequence, many of which contain coding regions, interspersed with blocks of satellite DNA.

Understanding the remarkable properties of heterochromatin will ultimately require physical mapping and genomic sequencing. The functional significance of heterochomatin remains obscure. Even closely related species can differ dramatically in the amount of heterochromatin and the nature of the satellite sequences (see, for example, John and Miklos[7]), and substantial increases or de-

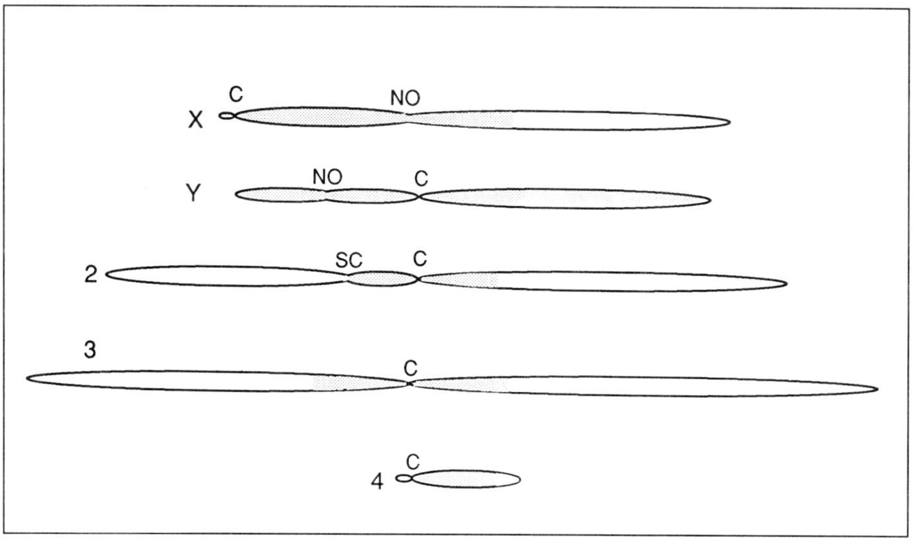

Fig. 3.1. Schematic representation of the chromosomes of D. melanogaster *as each appears in mitotic metaphase. The shaded parts denote constitutive heterochromatin. C, centromere; NO, nucleolus organizer; SC, prominent secondary constriction in 2L. Relative sizes of the chromatin blocks are taken from Ashburner M. Drosophila: A Laboratory Handbook, Cold Spring Harbor, NY: Cold Spring Harbor Laboratory Press 1993.*

creases in the quantity of heterochromatin appear to have little effect on developmental time, morphology, and viability of the organism,[41,42] at least under laboratory conditions. Fortunately, heterochromatic islands seem to have sufficient sequence complexity to build a physical map. P1 clones might be useful for this purpose because about 12.5% of the clones hybridize with the chromocenter formed from the heterochromatin in salivary-gland nuclei. On the other hand, P1 clones containing large blocks of simple satellite DNA have not been detected. Present information on the representation of heterochromatin and satellite DNA in the P1 and YAC libraries is discussed in detail in chapter 4. New approaches to the study of heterochromatin based on clones containing large fragments of DNA, as well as other genetic and cytological approaches discussed above, promise to illuminate the function and evolution of this enigmatic part of the genome.

MITOTIC AND POLYTENE CHROMOSOMES

The haploid chromosome set of *D. melanogaster* consists of two large metacentric autosomes (chromosomes 2 and 3), one small "dot" acrocentric autosome (chromosome 4), an acrocentric X chromosome (sometimes called chromosome 1), and a submetacentric Y chromosome. Each of the major chromosome arms (2L, 2R, 3L, 3R, and the X) has a length of about 2.5 μm, which may vary substantially depending on the tissue and method of preparation. Approximately one quarter of each large autosome is composed of a block of pericentric constitutive heterochromatin. Heterochromatin accounts for approximately half of the X chromosome and the entire Y chromosome, as well as most of the dot chromosome (Fig. 3.1).

Several larval and a few adult tissues of the Diptera undergo terminal differentiation accompanied by the development of giant polytene chromosomes. Chromosomes prepared from salivary glands of third instar larvae are usually the most suitable for cytological analysis. Each polytene chromosome results from approximately 10 rounds of DNA replication of the euchromatin without chromosome condensation or mitotic division. Much or all of the heterochromatin is not replicated or underreplicated. The euchromatic parts of the interphase sister chromatids come together and become precisely aligned longitudinally, like many parallel strands in a rope. The length and width of each polytene chromosome arm varies with species according to the extent of chromatin compaction and the number of rounds of replication (referred to as the level of polytenization). The length and width of the chromosome arms can also be strongly affected by technical procedures such as the method of fixation and squashing. In *D. melanogaster*, the length of each major chromosome arm in the polytene chromosomes is approximately 400 µm. Like the polytene chromosomes in other dipterans, those of *Drosophila* show a species-specific pattern of transverse bands resulting from variation in the local density of the chromatin along the euchromatic arms. The total number of bands is 5194, and their distribution among the chromosome arms is shown diagrammatically in Figure 3.2 along with the estimated total DNA content of the euchromatic bands in each arm. Note that only the euchromatin and part of the beta-heterochromatin undergoes polytenization. Most of the pericentromeric heterochromatin, including the entire Y chromosome, is underreplicated[5] and remains together to form an unstructured "cloud"

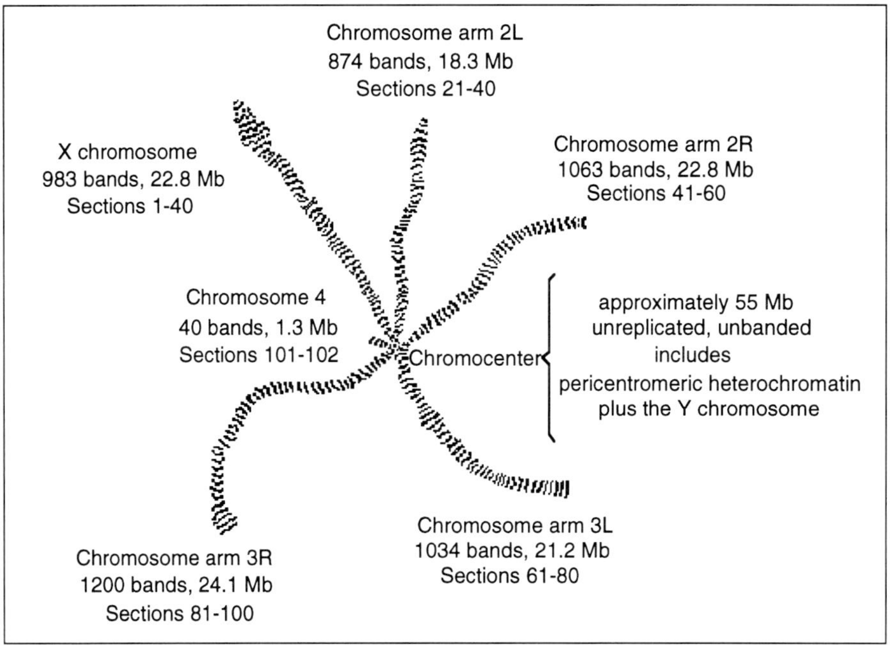

Fig. 3.2. A diagram of the general structure of squashed polytene chromosomes from nuclei of third instar larvae of D. melanogaster. *The sister chromatids of each chromosome are paired and longitudinally aligned forming the distinctive banding patterns in the euchromatin. For each chromosome arm, the number of bands and the estimated amount of euchromatic DNA is given. The constitutive heterochromatin remains underreplicated and forms a compact mass (the chromocenter) joining the bases of the arms.*

or compact mass called the chromocenter. Various aspects of the polytene chromosome structure have been reviewed by Lefevre[43] and Sorsa.[44]

The relatively high resolution afforded by in situ hybridization with polytene chromosomes allows rapid and efficient physical mapping of the euchromatic part of the *Drosophila* genome. This method, unfortunately, denies experimental access to heterochromatin except in providing important data about which clones apparently originate from these chromosome regions.

MOLECULAR ORGANIZATION OF THE *DROSOPHILA* GENOME

Reassociation kinetics indicates three major components of DNA sequences in the genome of *D. melanogaster.*[45] About 65% of the haploid DNA content (roughly 110 Mb) is represented by single copy sequences. Up to 12% of the genome (approximately 20 Mb) consists of a very heterogeneous, moderately reiterated, fraction. The remaining class of sequences (approximately 35 Mb) consists of relatively simple, highly repetitive, sequences.

DISPERSED MIDDLE REPETITIVE SEQUENCES

Middle repetitive sequences comprise nearly 20 Mb of the entire genome. About 5 Mb can be accounted for by the so-called localized repeats—the histone genes and the sequences coding for 5S RNA and rRNA. Much of the remaining 15 Mb is composed of multiple copies of each of a large number of different families of transposable elements, the genomic locations of which may differ within and between strains of flies. The number of copies of each family of elements varies from one organism to the next and between strains, but it is usually in the range between 10 and 100. The genetics and molecular biology of transposable elements have been thoroughly reviewed (see Berg and Howe,[46] for example) and for purposes of genome mapping requires only a brief overview.

Transposable elements may be divided into two major classes according to their mechanisms of transposition. One class includes elements that transpose by reverse transcription of an RNA intermediate. The other class includes elements that transpose through DNA. Within the RNA-intermediate class are two types:

1. Retrotransposons, which are elements flanked by long terminal repeats (LTRs), contain sequence elements that display similarities to retroviruses. Retrotransposons typically include two open reading frames (ORFs) corresponding to the retroviral genes *gag*, which codes for a precursor protein whose cleavage yields retroviral capsid constituents, and *pol*, which codes for a precursor protein also whose cleavage products include the reverse transcriptase and a protein implicated in proviral integration (integrase). The *pol* ORF thus includes domains corresponding to reverse transcriptase and integrase, as well as retroviral protease and RNase H. Among the major superfamilies of retroviruses are the *Ty1-copia* superfamily and the *Ty3-gypsy* superfamily, which are very widespread in invertebrate genomes and especially prominent in higher plants.[47,48] Members of the *Ty3-gypsy* superfamily encode a third product that appears to be a functional equivalent of the retroviral polypeptide coded by the *env* gene, which is the precursor of the surface glycoprotein. Presence of an *env*-like ORF in the *Ty3-gypsy* superfamily makes it even more similar to typical retroviruses. In *Drosophila*, the *gypsy* element itself has been shown to be infectious and can be transmitted horizontally.[49]
2. Nonviral retroposons, which are LINE-like elements lacking any terminal repeats and including a poly(dA)-rich sequence at the extreme 3' end; two ORFs are typically found, one of which encodes a sequence resembling reverse transcriptase. Elements of the LINE type in *Drosophila* include the *I* factor in *D. melanogaster*[50] and *Helena* in *D. virilis*.[51]

Another large, heterogeneous class of transposable elements are transposons that do not make use of an RNA intermediate in transposition. These mobile elements transpose directly from DNA to DNA. They include relatively short terminal inverted repeats and usually encode a special function (transposase) required for their own transposition. Examples in *Drosophila* include the *P* element of *D. melanogaster*,[52] used widely in germline transformation; *mari-*

ner,[53] found in *D. mauritiana* and large numbers of other insect and other invertebrate species;[54,55] and *Paris,* first identified in *D. virilis.*[51]

Transposable elements potentially have a profound role in shaping the eukaryotic genome. Gross chromosome rearrangements including translocations, deletions, and inversions can be created by events initiated by transposons.[56-59] Such rearrangements change the inheritance of groups of genes and can alter patterns of gene expression. Excisions and insertions of mobile elements can also affect the expression of nearby genes. Activation of various types of transposable elements by certain crosses leads to major genetic instabilities denoted by the term hybrid dysgenesis, which includes the induction of new mutations increasing genetic variability within and among populations. At the same time, the increased sterility of the hybrids has led to the suggestion that hybrid dysgenesis might, under some conditions, establish the initial reproductive isolation between sympatric or allopatric populations. In addition, transposable elements can affect the development and function of individual organisms.[60] For example, *P* element transposition has been demonstrated to increase the genetic variance of quantitative characters.[61]

On the other hand, the role of transposable elements in inducing mutation and genome rearrangements has been demonstrated almost exclusively in laboratory populations. Whether they play a similar role in nature has yet to be determined. It is possible that there is a natural selection for genetic modifiers that represses high rates of transposition in natural populations owing to its dysgenic effects. Repression of *P* element transposition is typical in natural populations of *D. melanogaster* owing to the presence of certain types of deletion derivatives of the elements themselves,[52] and the vast majority of *mariner* elements found in natural populations of insects have been mutationally inactivated.[54] Likewise, transposable elements are not prominent players in the scanty data that are presently available pertaining to the breakpoints of major chromosome rearrangements that are polymorphic within species or that differ between closely related species. For example, the breakpoints of the inversion In(3L)Payne, which is polymorphic virtually worldwide in populations of *D. melanogaster,* give no hint of having been generated by either directly by transpositional activity or by ectopic homologous recombination between different copies of the same transposable element.[62] The role of transposable elements in genome evolution is another example of an issue that can be addressed definitively with methods of physical mapping and genomic sequencing.

HIGHLY REPETITIVE, SIMPLE SEQUENCE, SATELLITE DNA

At least 21% of the haploid genome of *D. melanogaster* (approximately 35 Mb) consists of short, simple sequences repeated in tandem. These simple sequences are often referred to as satellite DNAs, and the extent of tandem repetition of the simple repeating unit can number in the millions. Some blocks of satellite DNA differ significantly in nucleotide composition from the rest of the genome which makes it possible to isolate these fractions as an additional band (a satellite band, hence the name) by centrifugation in a neutral CsCl gradient. Satellites are located preferably in the pericentromeric heterochromatin and arrayed in blocks of up to 750 kb. The organization of satellite sequences within the heterochromatin of different chromosomes has been examined and each chromosome has been found to have its own particular arrangements of satellites.

The satellites DNAs of *Drosophila* and their distribution among the chromosomes is discussed in some detail in chapter 4 in connection with screening the large-fragment DNA libraries with selected satellite sequences. Suffice it to say here that the major identified satellites in the genome of *D. melanogaster*, each accounting for more than 1% of the haploid genome, are 5'-AATAT-3' (3.1% of haploid genome), 5'-AAGAC-3' (2.4%), 5'-AAGAG-3' (5.6%), 5'-AAGAGAG-3' (1.5%), 5'-AATAACATAG-3' (2.1%), and a more-complex sequence with a repeating unit of 359 base pairs (5.1%).

The possible functions of satellite DNA remain a matter of intense speculations. Proposed roles include the protection of euchromatin, action as a mutagenic agent, and functions in chromosome recognition, pairing, and recombination. Much more research is needed to establish greater insight into possible functional and evolutionary roles of these ubiquitous and enigmatic sequences.

REFERENCES

1. Rubin GM. *Drosophila melanogaster* as an experimental organism. Science 1988; 240:1453-1459.
2. Merriam J, Ashburner M, Hartl DL, Kafatos FC. Toward cloning and mapping the genome of *Drosophila*. Science 1991; 254:221-225.
3. Ashburner M. *Drosophila: A Laboratory Handbook.* Cold Spring Harbor, NY: Cold Spring Harbor Laboratory Press 1989.
4. Bate M, Martinez Arias A, eds. *The Development of Drosophila melanogaster.* 2 Volumes. Cold Spring Harbor, NY: Cold Spring Harbor Laboratory Press 1993.
5. Rudkin GT. Non replicating DNA in *Drosophila.* Genetics (suppl. 1) 1969; 61:227-238.
6. Rasch EM, Barr HJ, Rasch RW. The DNA content of sperm of *Drosophila melanogaster.* Chromosoma 1971; 33:1-18.
7. John B, Miklos G. *The Eukaryote Genome In Development and Evolution.* London: Allen & Unwin 1988.
8. Muller HJ, Prokofieva AA. The individual gene in relation to the chromomere and the chromosome. Proc Nat Acad Sci USA 1935; 21:16-26.
9. Judd BH, Shen MW, Kaufman TC. The anatomy and function of a segment of the *X* chromosome of *Drosophila melanogaster.* Genetics 1971; 71:139-156.
10. Wilson R, Ainscough R, Anderson K, Baynes C, Berks M, Bonfield J, Burton J, Connell M, Copsley T, Cooper J, Coulson A, Craxton M, Dear S, Du Z, Durbin R, Favello A, Fraser A, Fulton L, Gardner A, Green P, Hawkins T, Hillier L, Jier M, Johnston L, Jones M, Kershaw J, Kirsten J, Laisster N, Latreille P, Lightning J, Lloyd C, Mortimer B, O'Callaghan M, Parsons J, Percy C, Rifken L, Roopra A, Saunders D, Shownkeen R, Sims M, Smaldon N, Smith A, Smith M, Sonnhammer E, Staden R, Sulston J, Thierry-Mieg J, Thomas K, Vaudin M, Vaughan K, Waterston R, Watson A, Weinstock L, Wilkinson-Sproat J, Wohldman P. 2.2 Mb of contiguous nucleotide sequence from chromosome III of *C. elegans.* Nature 1994; 368:32-38.
11. Levy LS, Manning JE. Messenger RNA sequence complexity and homology in developmental stages of *Drosophila.* Dev Biol 1981; 85:141-149.
12. Hall LM, Mason PJ, Spierer P. Transcripts, genes and bands in 315 000 base-pairs of *Drosophila* DNA. J Mol Biol 1984; 169:83-96.
13. Gausz J, Hall LMC, Spierer A, Spierer P. Molecular genetics of the *rosy-Ace* region of *Drosophila melanogaster.* Genetics 1986; 112:65-78.
14. Bossy B, Hall LMC, Spierer P. Genetic activity along 315 kb of the *Drosophila* genome. EMBO J 1984; 3:2537-2541.

15. Lowenhaupt K, Rich A, Pardue ML. Nonrandom distribution of long mono- and dinucleotide repeats in *Drosophila* chromosomes: correlations with dosage compensation, heterochromatin, and recombination. Mol Cell Biol 1989; 9:1173-1182.
16. Gatti M, Pimpinelli S, Santini G. Characterisation of *Drosophila* heterochromatin I. Staining and decondesation with Hoechst 33258 and quinacrine. Chromosoma 1976; 57:351-375.
17. Heitz E. Das heterochromatin der moose. Jahrb Wissenschaft Bot 1928; 69:762.
18. Barigozzi C, Dolfini S, Fraccaro M, Raimondi GR, Tiepolo L. *In vitro* study of the DNA replication patterns of somatic chromosomes of *Drosophila melanogaster.* Exp Cell Res 1966; 43:231-234.
19. Muller Hj, Painter T. The differentiation of the sex chromosomes of *Drosophila* into genetically active and inert regions. Z Indukt Abstammungs Vererbungsl 1932; 62:316-365.
20. Roberts PA. Difference in the behavior of eu- and heterochromatin: crossing-over. Nature 1965; 205:725-726.
21. Gall JG. Repetitive DNA in *Drosophila.* In: B. A. Hamkalo, J. Papaconstantinou, eds. *Molecular Cytogenetics.* New York: Plenum 1973; 59-73.
22. Miklos GLG, Cotsell JN. Chromosome structure at interfaces between major chromatin types: *alpha-* and *beta*-heterochromatin. BioEssays 1990; 12:1-6.
23. Hilliker AJ, Appels R, Schalet A. The genetic analysis of *D. melanogaster* heterochromatin. Cell 1980; 21:607-619.
24. Brosseau GE. Genetic analysis of the male fertility factors on the *Y* chromosome of *Drosophila melanogaster.* Genetics 1960; 45:257-274.
25. Kennison JA. The genetic and cytological organization of the *Y* chromosome of *Drosophila melanogaster.* Genetics 1981; 98:529-548.
26. Hazelrigg T, Fornili P, Kaufman TC. A cytogenetic analysis of X-ray induced male steriles on the *Y* chromosome of *Drosophila melanogaster.* Chromosoma 1982; 87:535-559.
27. Gatti M, Pimpinelli S. Cytological and genetic analysis of the *Y* chromosome of *Drosophila melanogaster.* Chromosoma 1983; 88:349-373.
28. Hilliker AJ. Genetic analysis of the centromeric heterochromatin of chromosome 2 of *Drosophila melanogaster*: deficiency mapping of EMS-induced lethal complementation groups. Genetics 1976; 83:765-782.
29. Hilliker AJ, Holm D. Genetic analysis of the proximal region of chromosome 2 of *Drosophila melanogaster.* I. Detachment products of compound autosomes. Genetics 1975; 81:705-721.
30. Marchant GE, Holm DG. Genetic analysis of the heterochromatin of chromosome 3 in *Drosophila melanogaster.* I. Products of compound-autosome detachment. Genetics 1988; 120:503-517.
31. Marchant GE, Holm DG. Genetic analysis of the heterochromatin

of chromosome 3 in *Drosophila melanogaster*. II. Vital loci identified through EMS mutagenesis. Genetics 1988; 120:519-532.

32. Baker WK. Position effect variegation. Adv Genet 1968; 14:133.
33. Spofford J. Position effect-variegation in *Drosophila*. In M. Ashburner, E. Novitski, eds. *Genetics and Biology of Drosophila*, Vol. 1c. London: Academic Press 1976; 955-1018.
34. Eissenberg JC, Morris GD, Reuter G, Hartnett T. The heterochromatin-associated protein HP-1 is an essential protein in *Drosophila* with dosage-dependent effects on position-effect variegation. Genetics 1992; 131:345-352.
35. Belyaeva ES, Demakova OV, Umbetova GH, Zhimulev IF. Cytogenetic and molecular aspects of position-effect variegation in *Drosophila melanogaster*. V. Heterochromatin-associated protein HP1 appears in euchromatic chromosomal regions that are inactivated as a result of position-effect variegation. Chromosoma 1993; 102: 583-590.
36. Clark RF, Elgin SC. Heterochromatin protein 1, a known suppressor of position-effect variegation, is highly conserved in *Drosophila*. Nucleic Acids Res 1992; 20:6067-6074.
37. Eberl DF, Duyf BJ, Hilliker AJ. The role of heterochromatin in the expression of a heterochromatic gene, the *rolled* locus of *Drosophila melanogaster*. Genetics 1993; 134:277-292.
38. Zhang P, Spradling AC. Insertional mutagenesis of *Drosophila* heterochromatin with single *P* elements. Proc Natl Acad Sci USA 1994; 91:3539-3543.
39. Karpen G, Murphy T, Le H. Molecular genetics of chromosome inheritance and heterochromatin in *Drosophila*. Abst 35th Ann *Drosophila* Res Conf 1994; 58.
40. Gonzalez C, Carmena M. Transposable elements are arranged in blocks that show a highly conserved pattern of distribution from β-heterochromatin to centromeres. Abst 35th Ann *Drosophila* Res Conf 1994; 323.
41. Miklos GL. Sequencing and manipulating highly repeated DNA. In G. A. Dover, R. B. Flavell, eds. *Genome Evolution*. London: Academic Press 1982; 41-68.
42. Miklos GLG. Localized highly repetitive DNA sequences in vertebrate and invertebrate genomes. In: R. J. MacIntyre, ed. *Molecular Evolutionary Genetics*. New York: Plenum 1985; 241-321.
43. Lefevre G Jr. A photographic representation and interpretation of the polytene chromosomes of *Drosophila melanogaster* salivary glands. In: M. Ashburner, E. Novitski, eds. *The Genetics and Biology of Drosophila*. London and New York: Academic Press 1976; 31-66.
44. Sorsa V. *Chromosome Maps of Drosophila*. 2 Volumes. Boca Raton, Florida: CRC Press 1988.
45. Crain WR, Eden FC, Pearson WR, Davidson EH, Britten RJ. Absence of short period interspersion of repetitive and nonrepetitive

sequences in the DNA of *Drosophila melanogaster*. Chromosoma 1976; 56:309-326.

46. Berg DE, Howe MM. *Mobile DNA*. Washington, D.C.: American Society for Microbiology 1989.
47. Flavell AJ. *Ty1-copia* group retrotransposons and the evolution of retroelements in the eukaryotes. Genetica 1992; 86:203-214.
48. Grandbastien M-A. Retroelements in higher plants. Trends Genet 1992; 8:103-108.
49. Kim A, Terzian C, Santamaria P, Pelisson A, Prud'homme N, Bucheton A. Retroviruses in invertebrates: The *gypsy* retrotransposon is apparently an infectious retrovirus of *Drosophila melanogaster*. Proc Natl Acad Sci USA 1994; 91:1285-1289.
50. Finnegan DJ. The I factor and I-R hybrid dysgenesis in *Drosophila melanogaster*. In: D. E. Berg, M. M. Howe, eds. *Mobile DNA*. Washington, D.C.: American Society for Microbiology 1989; 503-517.
51. Petrov D, Lozovskaya E, Schutzman J, Hartl D. Hybrid dysgenesis in *Drosophila virilis* mobilizes many transposable elements, one of which is shared with *Drosophila melanogaster*. Abst 35th Ann *Drosophila* Res Conf 1994; 3.
52. Engels WR. *P elements in Drosophila melanogaster*. In: D. E. Berg, M. M. Howe, eds. *Mobile DNA*. Washington, D.C.: American Society for Microbiology 1989; 437-484.
53. Hartl DL. Transposable element *mariner* in *Drosophila* species. In: D. E. Berg, M. Howe, eds. *Mobile DNA*. Washington, D.C.: American Society for Microbiology 1989; 531-536.
54. Robertson HM, MacLeod EG. Five major subfamilies of *mariner* transposable elements in insects, including the Mediterranean fruit fly, and related arthropods. Insect Mol Biol 1993; 2:125-139.
55. Robertson HM. The *mariner* transposable element is widespread in insects. Nature 1993; 362:241-245.
56. Schneuwly S, Kiroiwa A, Gehring WJ. Molecular analysis of the domonant homeotic *Antennapedia* phenotype. EMBO J 1987; 6:201-206.
57. Roeder GS, Fink GR. Transposable elements in yeast. In: J. A. Shapiro, ed. *Mobile Genetic Elements*. New York: Academic Press 1983; 299-326.
58. Engels WR, Preston CR. Identifying *P* factors in *Drosophila* by means of chromosome breakage hotspots. Cell 1981; 26:421-428.
59. Engels WR, Preston CR. Formation of chromosome rearrangements by *P* factor in *Drosophila*. Genetics 1984; 107:657-678.
60. Gvozdev VA, Belyaeva ES, Ilyin YV, Amosova IS, Kaidanov LZ. Selection and transposition of mobile dispersed genes in *Drosophila melanogaster*. Cold Spring Harbor Symp Quan Biol 1981; 45:673-685.
61. Mackay TFC. Transposable element-induced response to artificial selection in *Drosophila melanogaster*. Genetics 1985; 111:351-374.
62. Wesley CS, Eanes WF. Isolation and analysis of the breakpoint sequences of chromosome inversion *In(3L)Payne* in *Drosophila melanogaster*. Proc Natl Acad Sci USA 1994; 91:3132-3136.

CHAPTER 4

Randomness, Completeness, and Fidelity

The utility of a genomic library is determined not only by technical considerations such as ease of manipulation, size of insert, or yield of DNA, but also by considerations of quality including randomness, completeness, and fidelity. As applied to a library of clones, the term randomness refers to the representation of target genomic sequences. The target sequences are assumed to be small relative to the size of the cloned inserts. In a library that is random, each target sequence is present in a number of clones distributed according to a Poisson distribution with a mean equal to the number of copies of the target sequence in each haploid genome. Completeness refers to which sequences are present in the library. A library of clones is complete if every target sequence in the genome is present in the library. Because the sequences need not be represented in the library in proportion to the number of copies in the genome, a library may be complete but nonrandom. Furthermore, some target sequences may be missing by chance even in a random library because most collections of clones contain 5-10 haploid genome equivalents of DNA. Distinct from randomness and completeness is fidelity, which refers to the degree to which the genomic sequences replicated in clones are faithful copies of the target sequences present in the genome. Fidelity is an issue in regard to clones that are prone to rearrangement or clones containing chimeric inserts consisting of two or more sequences that are not contiguous in the genome.

For an assessment of the randomness, completeness, and fidelity of the *Drosophila* libraries, it is convenient to consider euchromatin separately from alpha and beta heterochromatin. Euchromatin constitutes the approximately 70% of the *Drosophila* genome that is highly replicated in the salivary gland nuclei and that forms the bands of the polytene chromosomes (chapter 3). Most but not all of the genes are located in euchromatin. The alpha heterochromatin contains long tracts of highly repeated, simple sequence, satellite DNA[1,2] that is located primarily in the pericentromeric regions of the autosomes, the base of the X chromosome, and the entire Y chromosome. In the salivary gland nuclei the alpha heterochromatin is grossly underreplicated relative to the euchromatin.[3] The beta heterochromatin contains repetitive sequences of greater complexity than those in the alpha heterochromatin; in the polytene chromosomes it is more or less underreplicated relative to the euchromatin and appears as a diffuse, meshlike region at the base of each chromosome arm plus the entire banded part of chromosome 4.[4]

In the following sections we present data indicating that the *Drosophila* P1 and YAC libraries are incomplete at least in regard to the highly repeated, simple sequence DNA characteristic of alpha heterochromatin. The libraries are apparently more complete,

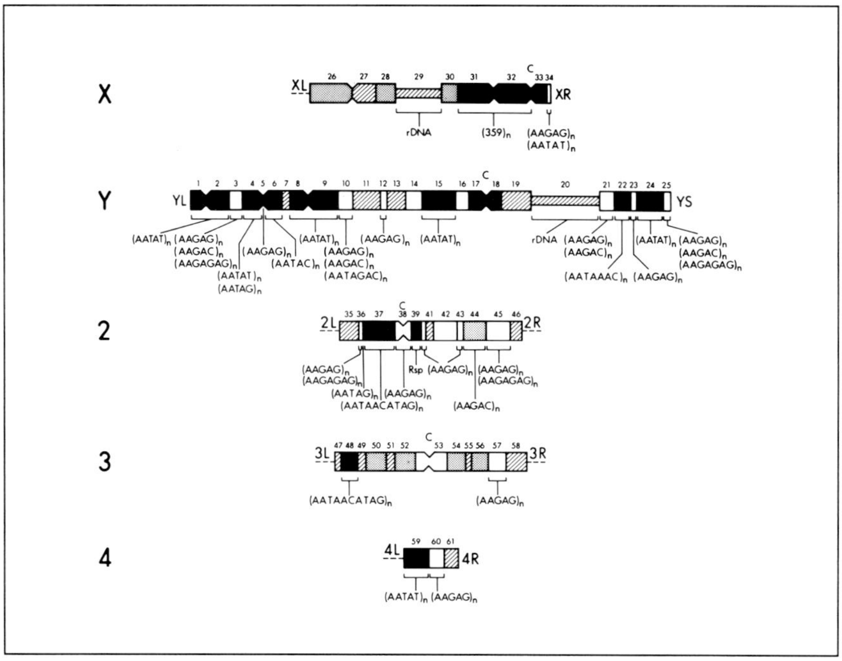

Fig. 4.1. Genomic organization and distribution of the major satellite DNA sequences in Drosophila. *From Lohe AR, Hilliker AJ, Roberts PA . Genetics 1993; 134:1149-1174.*

but very likely nonrandom, in regard to sequences found in the beta heterochromatin. Relative to euchromatic sequences, the data so far available support the hypothesis that the libraries are virtually complete and essentially random. Less information is available to address the issue of fidelity, and most of the evidence pertains to euchromatic sequences present in bacteriophage P1 clones. The primary observations are the sizes of PCR products of known genes and, to a more limited extent, direct DNA sequencing. However, both types of evidence suggest that most euchromatic P1 clones do faithfully replicate the *Drosophila* genomic sequences they contain.

HETEROCHROMATIN

Based on the relative lengths of the pericentromeric regions and of the Y chromosome in mitosis, it is generally thought that some 30-40% of the *Drosophila* genome consists of heterochromatin.[5] At least 21% of the genome consists of long tracts of tandemly repeated DNA sequences known as satellite DNA (Table 4.1).[1] The long tracts of satellite DNA, frequently interrupted by insertions of transposable elements and other more complex sequences, are located in the centromeric heterochromatin of chromosomes X, 2, 3, and 4; the Y chromosome consists almost entirely of such sequences (Fig. 4.1). The organization of the *Drosophila* genome is discussed further in chapter 3. The focus of the present discussion is the extent to which heterochromatic sequences are present in the P1 and YAC libraries.

HETEROCHROMATIN IN BACTERIOPHAGE P1 LIBRARIES

Among a total of 2,902 P1 clones localized by in situ hybridization, 364 clones (12.5%) yielded no major site of hybridization with the euchromatin but instead hybridized with the chromocenter. The majority (61%) of the clones hybridizing with the chromocenter also yielded multiple secondary sites of hybridization with individual bands dispersed in the euchromatin. This pattern of hybridization is assumed to result from moderately repetitive DNA sequences, primarily transposable elements, contained in P1 clones derived from the pericentromeric heterochromatin and the Y chromosome. The number of secondary sites of hybridization is typically in the range 25 to 100, but some clones yield as few as 5 and others as many as 200. The secondary sites of hybridization may result from two or more unrelated repetitive sequences present in the same P1 clone.

Is the finding that 12.5% of the clones hybridize with the chromocenter consistent with the composition of the *Drosophila* genome? As noted, at least 21% of the *Drosophila* genome consists of satellite DNA (Table 4.1).[1] In colony hybridizations, using as probes the simple repeating sequences $(AAGAC)_n$, which accounts for approximately 11% of *Drosophila* satellite DNA, and $(AATAT)_n$, which accounts for approximately 15% of the satellite DNA,[1] Allan Lohe has found that neither of the sequences is present in the P1 libraries (Lohe and Hartl, unpublished observations). Screening for the other simple satellite sequences has not been carried out but, given the results with $(AAGAC)_n$ and $(AATAT)_n$, it seems probable that other simple satellite sequences would also be absent from the P1 libraries. There are at least two reasons why satellites might be expected to be missing from the libraries or grossly underrepresented. First, the libraries were made from adult DNA, in which the satellite sequences are underrepresented (perhaps as much as ten-fold) relative to euchromatic sequences owing to the presence of polytene tissues, such as the Malpighian tubules, in the adult.[5] Even a small proportion of polytene cells contributes disproportionately to the deficiency of heterochromatin because, compared with a diploid cell, each polytene nucleus contributes up to 1,000-fold more to the single-copy DNA than it does to

***Table 4.1. Major satellite DNAs in* D. melanogaster**

Satellite sequence	Density[a]	Percent of genome[b]
5'-AATAT-3'	1.672	3.1 ± 0.4
5'-AATAG-3'	1.693	0.23 ± 0.02
5'-AATAC-3'	1.680	0.52 ± 0.04
5'-AAGAC-3'	1.689, 1.701	2.4 ± 0.2
5'-AAGAG-3'	1.705	5.6 ± 0.4
5'-AACAA-3'	1.663	0.06 ± 0.01
5'-AATAAAC-3'	1.669	0.23 ± 0.05
5'-AATAGAC-3'	1.688	0.24 ± 0.02
5'-AAGAGAG-3'	ND	1.5 ± 0.2
5'-AATAACATAG-3'	1.686	2.1 ± 0.2
359 bp repeat	1.688	5.1 ± 0.8

[a] Buoyant density in g/ml in CsCl [b] Mean ± standard error of the mean
Data from Lohe AR, Brutlag DL. Multiplicity of satellite DNA sequences in *Drosophila melanogaster*. Proc Natl Acad Sci USA 1986; 83:696-700.

the underreplicated heterochromatin. Secondly, long tracts of simple-sequence, tandemly repeated DNA promote intramolecular recombination, and hence clones containing satellite DNA are expected to be less stable genetically than clones containing nonrepeated sequences. In this context, it is worth emphasizing that some of the more complex satellites, such as the *Responder (Rsp)* component of segregation distortion, which is associated with a tandemly repeating unit of 116 base pairs in the centromeric heterochromatin of 2R,[6] is present in the P1 libraries; and P1 clones containing the *Bari* transposable element, located in the same region of 2R,[7] have also been recovered (Nurminsky and Hartl, unpublished observations).

Assuming, for the sake of argument, that the approximately 15% of the genome consisting of satellite sequences with simple repeating units ≤ 10 base pairs are not represented in the libraries, and assuming that the estimate of 30-40% is correct for the total amount of heterochromatin, one might expect 15-25% of the clones to hybridize with the chromocenter. The lower end of this range is not far from the observed proportion (12.5%) of clones that hybridize with the chromocenter. On the other hand, the apparently reasonable agreement with expectation may be coincidental. The estimates of the amount of heterochromatin and the amount of simple-sequence satellite DNA are indirect and very approximate, and so it is quite unwarranted to make categorical claims about the completeness of representation of heterochromatin in the libraries. All we can say for sure is that clones containing long tracts of $(AAGAC)_n$ and $(AATAT)_n$, and probably other simple-sequence satellite DNA, seem to be grossly underrepresented in the libraries. However, some clones containing complex repetitive sequences are present, including *Stellate* sequences (1.5 kb repeat) in the heterochromatin of the X chromosome[8] and the *Responder (Rsp)* component of segregation distortion[6] (116-bp repeat) in the heterochromatin of 2R (Lohe and Hartl, unpublished; Nurminsky and Hartl, unpublished).

HETEROCHROMATIN IN YAC LIBRARIES

The situation with regard to heterochromatic sequences in the YAC libraries appears to be similar to that in the P1 libraries. Among a total of 1,400 clones mapped by in situ hybridization,

201 hybridized primarily with the chromocenter and often gave evidence of multiple euchromatic sites of the type expected from the presence of transposable elements. The proportion of chromocentric YAC clones, 14.4%, is virtually identical with the 12.5% observed with in situ hybridizations of P1 clones.

The YAC libraries were screened with probes for $(AAGAC)_n$ and $(AATAT)_n$, which account for approximately 11% and 15% of *Drosophila* satellites, respectively; these probes are the same ones used to screen the P1 libraries. In addition, the YAC libraries were screened with probes for $(AAGAG)_n$ and $(AATAACATAG)_n$, which account for approximately 27% and 10% of *Drosophila* satellites, respectively.[1] The result of the satellite screening was no confirmed positives (Lohe and Hartl, unpublished observations). The YAC libraries therefore appear to lack clones containing long tracts of simple-sequence satellite DNA.

On the other hand, the YAC libraries do include clones containing more complex satellites. Colony hybridizations were carried out with a probe for a satellite with a repeating unit of 359 base pairs that has a buoyant density estimated as 1.688 g/cc.[9] This complex satellite is located primarily in a single block at the base of the X heterochromatin (Fig. 4.1). Among 3,072 YAC colonies screened, 10 contained the 359-bp satellite (Lohe and Hartl, unpublished observations); the 0.3% yield of clones is at least ten-fold smaller than the estimated amount (5%) of this satellite in the genome.[1] The shortage in the number of clones containing the 359-bp satellite may result from the fact that most of the YAC clones were prepared from DNA extracted from adults; alternatively, YAC clones containing highly repeated sequences might be somewhat unstable in yeast.[10]

The YAC libraries were also screened with other complex satellite sequences including *Responder (Rsp).* Positive clones were identified, including one *Not*I clone that contains most of the *Rsp* sequences present in the Oregon RC genome (Smoller and Hartl, unpublished observations). This clone hybridized with the chromocenter of the salivary gland nuclei and, in Southern blots, yielded a pattern of restriction fragments containing homology with *Rsp* indistinguishable from that observed in genomic digests. Further analysis of the clone indicated that its *Rsp* sequences occur in clusters separated by stretches of non*Rsp* DNA (Smoller and Hartl, unpublished).

Another family of repetitive sequences known to be present in the YAC libraries is the He-T family, which is located in heterochromatin and also near the telomeres.[11,12] In screening with a He-T clone designated Dm665, originally isolated because of its capacity for autonomous replication in yeast, approximately 20 clones were obtained from the *Not*I library.[13] In Southern blots, a number of these clones exhibited a restriction fragment homologous to Dm665 that is specific to the Y chromosome. This restriction fragment could also be demonstrated in genomic digests of females bearing particular portions of the Y chromosome from a series of X-Y translocations. Thus, the YAC libraries contain He-T sequences from the Y chromosome and these clones appear to be stable. Another set of Y-chromosome YACs has been isolated that contain the *Suppressor of Stellate [Su(Ste)]* locus,[14] which shares regions of homology with Dm665; many of the same YAC inserts contain both He-T and *Su(Ste)* sequences (Lohe and Hartl, unpublished).

The size distribution of YAC clones is not determined by physical requirements of the host-vector system such as the headful packaging mechanism in the construction of P1 clones. The unconstrained size distribution of YACs affords an unbiased test of whether YAC clones containing heterochromatin have the same size distribution as those containing euchromatin. For YACs that hybridize with the chromocenter, the mean size of the clone is 207.4 kb; for YAC clones that hybridize with the euchromatin, the mean size is 207.1 kb. Furthermore, the size distributions of the clones are virtually identical, which may be verified by comparing the distribution in Figure 4.2 with that of the euchromatic YACs illustrated in Figure 5.5.

RIBOSOMAL DNA

The ribosomal DNA in *D. melanogaster* is organized into tandem arrays of repeating sequences ranging in length from 11-17 kb.[15] The rDNA arrays are cytologically visible as nucleolus organizers. On average there are approximately 200 copies (2,800 kb) in the centromeric heterochromatin of the X chromosome and approximately 150 copies (2,200 kb) in the short arm of the Y chromosome.[5] However, there is substantial variation in the number of rDNA copies from one fly stock to the next. For example, the X chromosomes in different wildtype stocks may differ up to five-fold in the number of rDNA copies they contain.[16] On average, assuming a total genome size of 165,000 kb, the amount of rDNA present in a *Drosophila* genomic library is expected to be approximately 1.6%.[5]

The organization of the rDNA repeating unit is illustrated in Figure 4.3. The functional unit produces a single transcript that is processed to yield the mature 18S, 28S, 5.8S, and 2S RNA molecules present in the ribosome. In genomic DNA, the repeating units are separated by nontranscribed and transcribed spacer sequences of variable length, and the rDNA coding sequences are separated by transcribed spacers also variable in length. The unit repeats in the rDNA array are apparently not all in the same orientation, but the number and distribution of polarity reversals along the array has not been well characterized and may differ from one strain to the next.[15]

Many of the 28S coding sequences are interrupted by unrelated retrotransposon-like elements lacking long terminal repeats

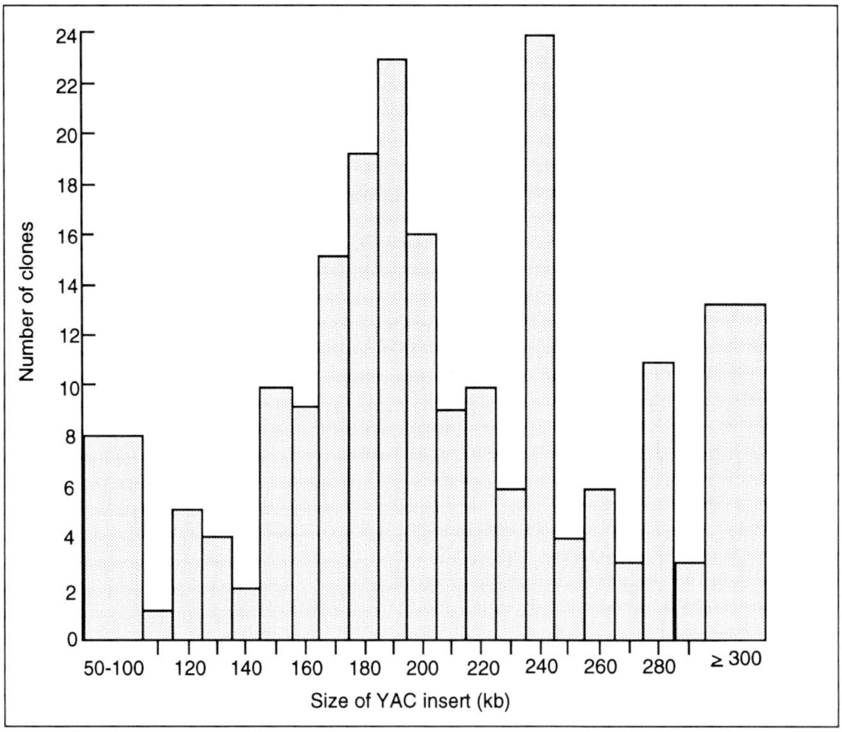

Fig. 4.2. Distribution of size of insert in YAC clones localized by in situ hybridization to the chromocenter.

denoted R1 (also called Type I) and R2 (also called Type II). The full-length R1 and R2 elements are 5.3 kb and 3.6 kb in length, respectively, but there are many copies present in the 28S rDNA that are truncated to different extents at the 5' end.[16] The insertion sites of R1 and R2 are separated by 74 base pairs (Fig. 4.3), and a particular 28S gene can contain no insertion, R1 only, R2 only, or both R1 and R2. In one study, among *Drosophila* strains from diverse geographic locations, the proportion of 28S genes with R1 inserts ranged from 17% to 67% (average 38%) and the proportion with R2 inserts ranged from 2% to 28% (average 16%).[16] The presence of R1 and R2 insertions in the 28S gene are statistically independent, and so the probability of both insertions in a particular 28S gene is very nearly the product of the probability of the single insertions.[16] The functional effects of R1 and R2 insertion are profound because either insertion eliminates or severely inhibits rDNA transcription.[17]

RIBOSOMAL DNA IN BACTERIOPHAGE P1 LIBRARIES

The P1 libraries were screened with a 1.3 kb probe for 28S rRNA in order to ascertain the ability of the P1 cloning system to clone this type of complex repeated DNA sequence. Among 3,840 P1 clones in the pNS582-tet14 Ad10 vector, 64 clones were positive. The proportion of positive clones (1.7%) is a good fit with the expected amount of rDNA in the genome of an average *Drosophila* strain, although the abundance of rDNA in the *y; cn bw sp* stock used in construction of the P1 libraries was not determined specifically. As expected, P1 clones containing rDNA hybridize with the chromocenter upon in situ hybridization.

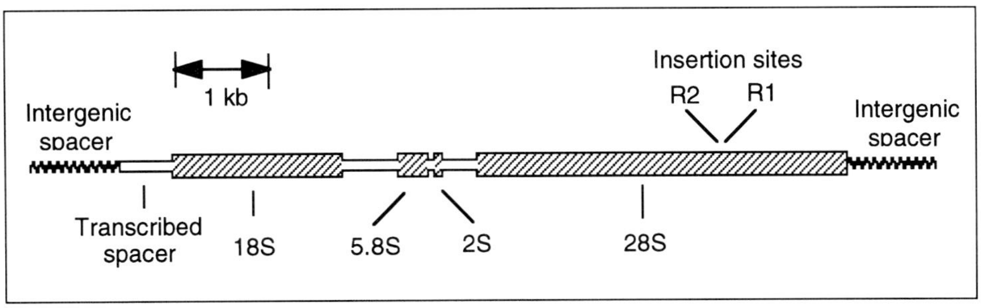

Fig. 4.3. Organization of the rDNA repeating unit in Drosophila. *Data from Jakubczak JL, Zenni MK, Woodruff RC, Eickbush TH. Genetics 1992; 131:129-142.*

The pNS582-tet14 Ad10 library was also screened with a probe for the spacer (average size 240 bp) between rDNA repeats (Fig. 3). Among 70 positive clones isolated, 64 coincided with those containing 28S rDNA identified previously. The additional six clones containing the rDNA spacer had no detectable 28S rDNA.

To investigate the stability of P1 clones containing 28S rDNA, eight clones were analyzed in additional detail. The average size of the inserts was 79 kb, approximately the same size as P1 clones containing nonrepetitive sequences derived from the euchromatin, suggesting that no large deletions had occurred during construction of the rDNA clones or during their propagation in *E. coli*. The viability and stability of two of the rDNA-containing clones was examined by repeatedly subculturing a single colony of each clone overnight in liquid culture and reisolating a single colony. The subculturing and overnight growth was repeated four times. After each overnight growth cycle, the single colony isolated was saved for DNA isolation and analysis. Restriction digests of the repeated subcultures gave no evidence of deletions or complex rearrangements of the rDNA-containing P1 clones.[18] These observations provide additional support for the view that the P1 system is capable of the cloning and faithful replication of repetitive DNA sequences provided that the repeating units are of sufficient sequence complex.

RIBOSOMAL DNA IN YAC LIBRARIES

Screening for 28S rDNA genes was also carried out among 768 clones in the pYACP-1 library described in chapter 2, which contains inserts of randomly sheared DNA from the Oregon RC stock.[19] A total of 38 clones containing 28S rDNA were isolated, which represents 4.9% of all the clones screened. Among the 38 rDNA clones, 27 (71%) were also positive when assayed for R1 inserts and 17 (45%) were positive when assayed for R2 inserts. These numbers are difficult to interpret in terms of R1 and R2 occupancy of the 28S rDNA genes because each pYACP-1 clone contains multiple rDNA repeats and because the distribution of R1 and/or R2 in the rDNA may be somewhat clustered. Nevertheless, it is clear that many of the 28S copies in the pYACP-1 clones contain R1 and/or R2 insertions. It should also be noted that one clone was isolated that contains the R1 sequence in the absence of rDNA. Presence of the R1 element in centromeric heterochromatin apart from the rDNA cistrons has been observed previously.[16,20]

At least four of the rDNA clones were obviously genetically unstable in yeast as judged by the finding of multiple YAC bands in one or more of the subcultures derived from single yeast colony replicates. The observed frequency of 10% of unstable clones should be considered a minimum estimate because a greater frequency of unstable clones might very well be observed in a more systematic or larger-scale study. In contrast with the rDNA clones, YAC clones containing other types of complex repetitive DNA sequences, such as *Responder* and *Suppressor of Stellate*, appear to be genetically quite stable in yeast.

TOO FEW CLONES FROM CHROMOSOME 4

Tiny chromosome 4 affords an interesting test of the randomness of recovery of clones containing beta heterochromatin and other types of complex repetitive sequences because the banded part of chromosome 4 contains such sequences in abundance and is regarded as consisting entirely of beta heterochromatin.[4,21] The data are very clear: both the P1 libraries and the YAC libraries contain far fewer clones from chromosome 4 than expected. The nonheterochromatic part of chromosome 4 is about 1 Mb in length, or approximately 1% of the entire euchromatic genome. However, among 2,461 P1 clones that were localized to single major sites of hybridization in the euchromatin, only 7 clones hybridized with chromosome 4.[22] The expected number of clones hybridizing with chromosome 4 is about 25, and the discrepancy is statistically significant ($P < 0.01$). In the case of YAC clones, only three among 1193 euchromatic clones hybridized with chromosome 4 as compared with an expected number of 11.9 ($P = 0.01$).[23]

Overall, the percentage of chromosome 4 clones in the P1 libraries (0.28%) is virtually identical to that in the YAC libraries (0.25%), and in both cases the underrepresentation is between 70% and 80%. The clear implication is that chromosome 4 is not present in either set of libraries in proportion to its representation in the genome. It may even be the case that some regions of chromosome 4 are not present at all.

EUCHROMATIC SEQUENCES

Perhaps the toughest challenge for any cloning system is the stable propagation of repetitive DNA. It was primarily to evaluate the adequacy and reliability of the P1 and YAC systems in accommodating these types of sequences that we investigated the presence and stability of repetitive and other heterochromatic sequences in the clones. However, most of the genetic activity in *Drosophila* is in the euchromatin; that is where most of the genes are: among 3,758 *Drosophila* genes tabulated by chromosomal location,[24] only a handful are in heterochromatin. Examples of genes in heterochromatin include *light*[25] and *rolled*[26] in 2L and 2R, respectively, as well as the fertility factors in the Y chromosome.

The final word on the fidelity of P1 and YAC clones must await large-scale DNA sequencing of P1 clones and further analysis of the YACs. The limited data available at this time were collected primarily to determine whether the cloning of euchromatic sequences was grossly nonrandom and whether a large fraction of the clones might undergo major rearrangements in the host cells. It is encouraging that these data, as well as the in situ hybridization data discussed in the next chapter, suggest that the cloning of euchromatic sequences, although perhaps not completely random in a statistical sense, is nevertheless sufficiently representative that a very large fraction of the genome can be expected to be present in the P1 and YAC libraries. Furthermore, there is thus far no evidence that major rearrangements take place at a high frequency in most clones.

REPRESENTATION OF EUCHROMATIC SEQUENCES IN BACTERIOPHAGE P1 LIBRARIES

With regard to euchromatic sequences, the P1 libraries were initially examined for the presence and fidelity of replication of sequences corresponding to a number of known single-copy genes as well as to a number of anonymous cDNAs.[18] P1 clones containing these sequences were identified by colony hybridization of the pNS582-tet14Ad10 library consisting of 3,840 clones. The initial screening was carried out with probes for the genes *alcohol dehydrogenase, rosy*, and *odd-skipped* as well as with probes derived from 17 different eye-specific cDNAs taken from a sorted cDNA library.[27] The numbers of clones recovered using these probes were not significantly different from those expected assuming randomness of representation and completeness of coverage of the euchromatin. Among the 3,840 clones, about 85% (3,264) are expected to derive from euchromatin and, assuming an average insert size of 80 kb, the total euchromatic DNA content among the clones is 260 Mb. Assuming a euchromatic genome size of 110 Mb, the number of haploid genomes in the euchromatic clones is approximately 2.4. The expected average number of positive clones per probe is also 2.4, which compares quite favorably with the observed average of 2.0. Furthermore, the Poisson expectation among 20 probes is 1.8 with no positives, 4.4 with one, and 13.8 with two or more. The observed results were 0 with no positives, 9 with one, and 11 with two or more. Although the results fit satisfactorily in a statistical sense, there is a tendency for too many probes to be recovered in only one clone, which may suggest some nonrandomness of coverage. Alternatively, perhaps the euchromatic genome is somewhat greater than 110 Mb or perhaps fewer than 85% of the clones in the pNS582-tet14Ad10 library derive from euchromatin.

To test for the fidelity of the P1 cloning system, detailed tests were carried out with all 39 P1 clones recovered from the 20 probes. Southern blots were carried out with each probe to determine whether the clones contained restriction fragments coincident in size with those from genomic DNA. In each case, one or more diagnostic restriction fragments were observed in genomic DNA that proved to be identical in size to restriction fragments obtained from the P1 clones (Petrov and Hartl, unpublished).[18] Furthermore, different P1 clones obtained by hybridization with the same single-copy probe gave identical patterns of diagnostic restriction fragments in Southern blots when hybridized with the probe. Also important in regard to the use of P1 clones for physical mapping, different P1 clones recovered with the same probe hybridized to the same or overlapping sets of bands in the salivary gland chromosomes.

Strongly encouraged by the initial tests suggesting virtually complete and nearly Poisson coverage of the euchromatin in the pNS582-tet14Ad10 library, additional libraries were constructed in pAd10 sacBII and the extensive series of in situ hybridizations began. These studies began in July, 1991, and were carried out primarily by Robert W. Jones and E. R. Lozovskaya.

Much more data regarding coverage of the euchromatin in the P1 libraries, as well as the fidelity of replication of P1 clones, has emerged from STS screening of the libraries in order to assemble sets of P1 clones inferred to overlap at the molecular level by virtue of their ability to support PCR amplification of single-copy genomic sequences. These studies were described briefly in chapter 2 and the details will be provided in chapter 6. In addition to contig assembly, the STS screening also provides data regarding library coverage and fidelity. Coverage is addressed by the number of single-copy sequences that yield no positive clones upon PCR screening. Although false negatives do occasionally happen in PCR screening, most of the STSs yielding no P1 clones are probably missing from the libraries. The results of PCR screening for genes whose sequences are present in GenBank are shown in Table 4.2. The X chromosomal genes are tabulated separately from the autosomal genes because, for reasons explained below, the X chromosome is underrepresented in the P1 libraries.

Considering first the autosomal genes, among 9,216 P1 clones screened, the average number of positive clones per probe is 3.7. Compared with the Poisson distribution, there are too many STSs with no representatives in the library (observed 9, expected 4.8) and too many with multiple hits (observed 15, expected 7.2). The excess number of unrepresented clones may result from screening for STSs in some regions that are difficult to clone that are underrepresented in the library or from false negatives in the PCR reactions. The excess of multiple hits may result from certain clones in the library that were inadvertently duplicated in picking the initial colonies or from STSs that support amplification of a PCR product of the same size from different members of a gene family (for example, transmembrane receptors).

The X chromosome must be considered separately for two reasons. First, the representation of the X chromosome in the library is expected to be 3/4 that of the autosomes because the clones were made from DNA extracted from males and females together. Because females have two X chromosomes and males have one—whereas each sex has two copies of each autosome—the ratio of X DNA to autosomal DNA in the libraries is expected to be 3 : 4. The second complication is that the libraries were made from DNA extracted from flies taken from culture bottles that had been maintained for various lengths of time. Because *Drosophila* males generally live longer than females, the X chromosome could well be underrepresented in the libraries by even more than 25%.

The X-chromosomal data in Table 4.2 clearly indicate that X chromosomal genes are significantly underrepresented relative to autosomal genes. The average number of positive clones for X-linked STSs is 1.8 which is approximately half that observed in the case of autosomal STSs. Evidently, the population of adult flies used for the DNA preparations had a very large excess of males. The X-linked STS data also show departures from Poisson expectation in the same direction as that observed for autosomal STSs. In order to compensate for the underrepresentation of the X chromosome, the X-linked STSs are also being isolated from DNA pools from an additional 9,216 P1 clones in pAd10 sacBII that had been isolated and maintained as a backup.

The fidelity of P1 replication is addressed in the STS screening by the size of the observed PCR product. The oligonucleotides used for screening are typically designed to amplify a genomic fragment of approximately 300 to 500 base pairs, and the amplification of a single fragment of the correct size is confirmed by PCR amplification of total genomic DNA prior to the library screening. In screening the P1 libraries, any clone that has a deletion of a significant part of the sequence between the primers will be detected because PCR amplification will yield a single fragment that is unexpectedly small in size. Among the STSs in Table 4.2, clones containing such putative deletions have not been detected.

Table 4.2. Number of positive clones per known gene in P1 libraries

Number of positive clones	X-linked STS	Autosomal STS
0	23	9
1	19	29
2	20	33
3	7	36
4	11	26
5	5	20
6	1	22
7	0	12
8	0	8
9	0	2
≥10	0	5

In addition to the STS screening with known genes, extensive screening is also currently being carried out with sequences determined from the ends of the inserts in P1 clones previously mapped by in situ hybridization. These results are somewhat biased in regard to the completeness of representation in the P1 libraries because the sequences obtained from inserts in P1 clones must perforce be present in clonable regions. However, the screening is not biased in regard to the detection of putative deletions, and again no significant number of putative deletions has been detected.

It should also be noted that, by the Fall of 1994, several P1 clones with inserts from the *Bithorax* complex, the *Antennapedia* complex, and the *alcohol dehydrogenase* region had either been sequenced in their entirely or were in the process of being sequenced at the Lawrence Berkeley Laboratories of the *Drosophila* Genome Center. These studies also yielded no evidence for instability or rearrangement of euchromatic sequences maintained in the P1 clones.

REPRESENTATION OF EUCHROMATIC SEQUENCES IN YAC LIBRARIES

Less data are available regarding the completeness and fidelity of the YAC libraries than of the P1 libraries because the P1 clones have assumed primacy in contig assembly and DNA sequencing. Judging from the results of the in situ hybridizations discussed in chapter 5, however, the coverage of the euchromatin in the YAC libraries appears to be widespread and virtually complete. There is a deficiency of clones from chromosome 4, as noted above. There is also the expected underrepresentation of clones from the X chromosome resulting from the fact that the libraries were constructed from a mixture of male and female DNA.[28] Specifically, among 1190 euchromatic YAC clones, 184 hybridized with the X chromosome and 1,006 to either of the large autosomes. The average number of clones per autosomal arm is therefore 1,006/4 = 251.5; because the X chromosome is approximately the size of one of the autosomal arms, the ratio of X-chromosomal clones to autosomal clones is 184/251.5 = 73%, which is very close to the expected 75%. Virtually complete coverage of the euchromatin in the YAC libraries is also supported by the sizes of the gaps remaining in the framework YAC physical map discussed in chapter 5, which are of the number and size that would be expected statistically given the insert size and number of clones mapped.[29]

In regard to the fidelity of replication of YAC clones, the initial tests were also quite encouraging. Tests of completeness and fidelity of the YAC cloning system were carried out initially by screening the libraries with probes for single-copy genes. These included probes for the genes *white, rosy, alcohol dehydrogenase,* and *abdominal-A.* Each probe identified at least one YAC clone that was subsequently confirmed by Southern hybridization of field inversion gels probed with the appropriate sequence as well as by direct in situ hybridization with the polytene chromosomes. The internal structures of the YAC clones containing *white, rosy,* and *alcohol dehydrogenase* were examined by comparing the lengths of restriction fragments from the YAC clones with those in the Oregon-R genome.[19] Restriction fragments detected in Southern blots using each of the YAC clones were of the expected size and matched equivalent bands in the Oregon R genome. A detailed analysis of 160 kb of cloned DNA derived from the *Bithorax* complex yielded similar results (Garza and Hartl, unpublished), as did a series of clones from the *Antennapedia* complex.[28,30] In addition, each of the euchromatic YAC clones appeared to be genetically stable in the sense that they did not appear to rearrange at frequencies high enough to be detected during routine subcloning in yeast. On the other hand, detailed tests have been carried out on a small scale with a relatively small number of clones, and we cannot exclude the possibility that some fraction of the euchromatic YAC clones may be rearranged or unstable.

References

1. Lohe AR, Brutlag DL. Multiplicity of satellite DNA sequences in *Drosophila melanogaster.* Proc Natl Acad Sci USA 1986; 83:696-700.
2. Lohe AR, Hilliker AJ, Roberts PA. Mapping simple repeated DNA sequences in heterochromatin of *Drosophila melanogaster.* Genetics 1993; 134:1149-1174.
3. Rudkin GT. Non replicating DNA in *Drosophila.* Genetics 1969; 61 (suppl. 1):227-238.
4. Miklos GLG, Cotsell JN. Chromosome structure at interfaces between major chromatin types: *alpha-* and *beta-*heterochromatin. BioEssays 1990; 12:1-6.
5. Ashburner M. *Drosophila: A Laboratory Handbook.* Cold Spring Harbor, NY: Cold Spring Harbor Laboratory Press 1989.
6. Wu C-I, Lyttle TW, Wu M-L, Lin G-F. Association between a satellite DNA sequence and the *Responder* of *Segregation Distorter* in *D. melanogaster.* Cell 1988; 54:179-189.

7. Caizzi R, Caggese C, Pimpinelli S. *Bari-1*, a new transposon-like family in *Drosophila melanogaster* with a unique heterochromatin organization. Genetics 1993; 133:335-345.
8. Shevelyov YY. Copies of a *Stellate* gene variant are located in the X heterochromatin of *Drosophila melanogaster* and are probably expressed. Genetics 1992; 132:1033-1037.
9. Hsieh TS, Brutlag D. Sequence and sequence variation within the 1.688 g/cm^3 satellite DNA of *Drosophila melanogaster*. J Mol Biol 1979; 135:465-481.
10. Brownstein BH, Nelson DL, eds. *YAC Libraries: A User's Guide*. New York, W. H. Freeman 1994.
11. Traverse KL, Pardue ML. Studies of He-T DNA sequences in the pericentric regions of *Drosophila* chromosomes. Chromosoma 1989; 97:261-271.
12. Levis RW, Ganesan R, Houtchens K, Tolar LA, Sheen F. Transposons in place of telomeric repeats at a *Drosophila* telomere. Cell 1993; 75:1083-1093.
13. Danilevskaya ON, Kurenova EV, Pavlova MN, Link AJ, Koga A, Vellek A, Hartl DL. He-T family DNA sequences in the Y chromosome of *Drosophila melanogaster* share homology with the X-linked Stellate genes. Chromosoma 1991; 100:118-124.
14. Livak KJ. Detailed structure of the *Drosophila melanogaster Stellate* genes and their transcripts. Genetics 1990; 124:303-316.
15. Williams SM, Robbins LG. Molecular genetic analysis of *Drosophila* rDNA arrays. Trends Genet 1992; 8:335-340.
16. Jakubczak JL, Zenni MK, Woodruff RC, Eickbush TH. Turnover of R1 (Type I) and R2 (Type II) retrotransposable elements on the ribosomal DNA of *Drosophila melanogaster*. Genetics 1992; 131:129-142.
17. Jamrich M, Miller OL. The rare transcripts of interrupted rDNA genes in *Drosophila melanogaster* are processed or degraded during synthesis. EMBO J 1984; 3:1541-1545.
18. Smoller DA, Petrov D, Hartl DL. Characterization of bacteriophage P1 library containing inserts of *Drosophila* DNA of 75-100 kilobase pairs. Chromosoma 1991; 100:487-494.
19. Garza D, Ajioka JW, Burke DT, Hartl DL. Mapping the *Drosophila* genome with yeast artificial chromosomes. Science 1989; 246:641-646.
20. Peacock WJ, Appels R, Endow S, Glover D. Chromosomal distribution of the major insert in *Drosophila melanogaster* 28S rRNA genes. Genet Res 1981; 37:209-214.
21. Miklos GLG, Yamamoto M, Davies J, Pirrotta V. Microcloning reveals a high frequency of repetitive sequences characteristic of chromosome 4 and the β-heterochromatin of *Drosophila melanogaster*. Proc Natl Acad Sci USA 1988; 85:2051-2055.

22. Hartl DL, Nurminsky DI, Jones RW, Lozovskaya ER. Genome structure and evolution in *Drosophila:* Applications of the framework P1 map. Proc Natl Acad Sci USA 1994; 91:6824-6829.
23. Hartl DL. Genome map of *Drosophila melanogaster* based on yeast artificial chromosomes. In K. E. Davies, S. M. Tilghman, eds. *Genome Analysis Volume 4: Strategies for Physical Mapping.* Cold Spring Harbor, NY: Cold Spring Harbor Laboratory Press 1992; 39-69.
24. Merriam J, Ashburner M, Hartl DL, Kafatos FC. Toward cloning and mapping the genome of *Drosophila.* Science 1991; 254:221-225.
25. Devlin RH, Bingham B, Wakimoto BT. The organization and expression of the *light* gene, a heterochromatic gene of *Drosophila melanogaster.* Genetics 1990; 125: 129-140.
26. Biggs WHI, Zavitz KH, Dickson B, van der Straten A, Brunner D, Hafen E, Zipursky SL. The *Drosophila rolled* locus encodes a MAP kinase required in the sevenless signal transduction pathway. EMBO J 1994; 13:1628-1635.
27. Palazzolo MJ, Hyde DR, VijayRaghavan K, Mecklenberg K, Benzer S, Meyerowitz E. Use of a new strategy to isolate and characterize 436 *Drosophila* cDNA clones corresponding to RNAs detected in adult heads but not early embryos. Neuron 1989; 3:527-539.
28. Ajioka JW, Smoller DA, Jones RW, Carulli JP, Vellek AEC, Garza D, Link AJ, Duncan IW, Hartl DL. *Drosophila* genome project: One-hit coverage in yeast artificial chromosomes. Chromosoma 1991; 100:495-509.
29. Hartl DL, Palazzolo MJ. *Drosophila* as a model organism in genome analysis. In K. W. Adolph, eds. *Genome Research in Molecular Medicine and Virology.* Orlando, FL: Academic Press 1993; 115-129.
30. Garza D, Ajioka JW, Carulli JP, Jones RW, Johnson DH, Hartl DL. Physical mapping of complex genomes. Nature 1989; 340:577-578.

CHAPTER 5

COVERAGE OF THE EUCHROMATIC GENOME

The initial goal of the *Drosophila* genome project was to create a physical map—a linear representation of molecular landmarks (typically the locations of DNA restriction sites, probes, or sequences) in which distances are ideally scaled in proportion to the number of nucleotides. In contrast, the traditional linkage maps of genetic analysis have distances that are scaled in proportion to the amount of recombination.[1] The two kinds of maps are not uniformly congruent because recombination rates are not uniform along a DNA molecule. The correlation of the physical map with the linkage map constitutes a genome map, in which the scale is proportional to physical distance but the landmarks are the locations of genes.

The earliest physical mapping studies employed a bottom-up strategy,[2-4] in which clones containing relatively small inserts of DNA from the target genome were analyzed by restriction mapping or fingerprinting, and overlapping clones were arranged in contiguous units (contigs) through the recognition of DNA fragments that pairs of clones had in common. An alternative strategy, called a top-down strategy, relies on the larger DNA inserts that can be cloned in P1 or YAC vectors.[5-7] Both strategies require molecular techniques to assemble clones into contigs. Typically, the average contig size is a small multiple of the size of the inserts in the clones. Once contigs of appreciable size have been assembled, they can be correlated with the linkage map. Singleton clones and unmapped contigs are assigned to a sort of clone limbo where their positions in the genome map, and their relationships to one another, are indeterminate.

The special organizing feature that *Drosophila* presents for physical mapping is the set of bands in the polytene chromosomes in the salivary glands. Using the technique of in situ hybridization, large numbers of clones can be assigned locations in the genome relatively rapidly, forming a framework map in which the approximate locations of the clones are determined independently and without regard to molecular overlap with other clones.[8] Subject to the limit of resolution of the cytological procedures and the reliability of the assigned locations, the clones in the framework map can also be analyzed to estimate overall coverage in terms of gaps as well as to identify regions across which there are apparently overlapping clones as judged cytologically.[9,10]

The size of the *Drosophila* genome is estimated as 165 Mb (millions of base pairs).[11,12] Assuming that the euchromatin accounts for two-thirds of the genome, the amount of genomic DNA represented in the polytene arms is 110 Mb. This estimate is strongly influenced by the amount of heterochromatin that is assumed: if the assumption were that the amount of heterochromatin ranged from 25% to 40%, then the estimated size of the euchromatic genome would range approximately from 100 Mb to 125 Mb. The

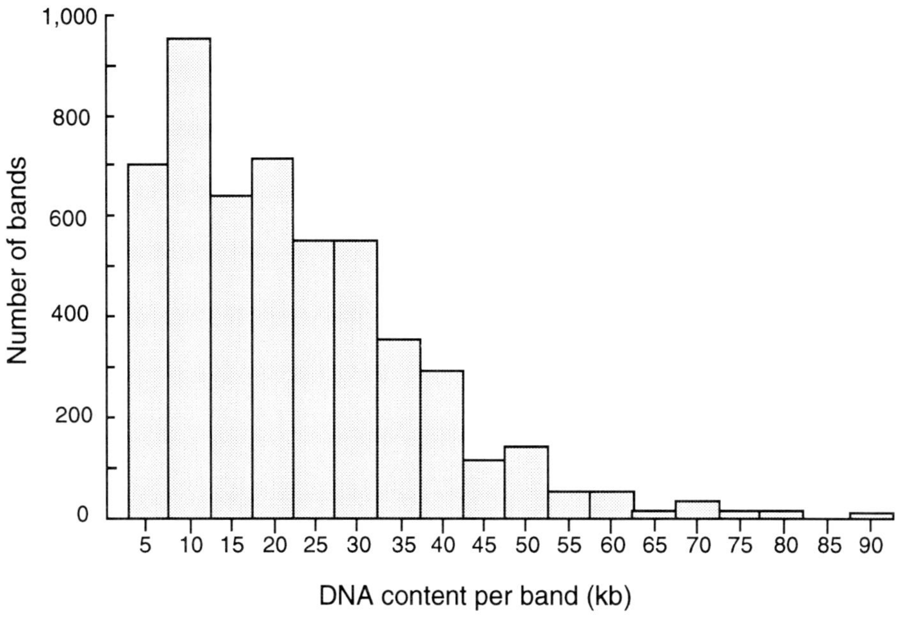

Fig. 5.1. Distribution of estimated DNA content of bands in the polytene salivary gland chromosomes. Data from Heino TI, Saura AO, Sorsa V. Drosophila *Information Service 1994; 73:619-738.*

classical maps of the polytene salivary gland chromosomes developed by Calvin Bridges and his son, Philip, depict about 5,000 bands.[13-18] More recent studies with electron microscopy reveal 5,194 bands distributed among the chromosomes as shown in Table 5.1.[19] Also shown in Table 5.1 are estimates of DNA content per chromosome arm. These were obtained by summing, for each arm, the DNA content per band as estimated by Heino et al[19] from band thickness and density in sections prepared for electron microscopy. The estimates of DNA content per band were normalized to a total euchromatic DNA content of approximately 110 Mb.[19] Using this estimate, the average polytene band represents a little over 21 kb of genomic DNA. The average of 21 kb conceals a wide variation in DNA content per band. As shown in Figure 5.1, the distribution of DNA content per band ranges from 5 kb to 90 kb and has a standard deviation of 14 kb.

Through in situ hybridization with the polytene chromosomes, the approximate locations of clones covering much of the genome can be assembled relatively rapidly because cloned DNA sequences can be assigned positions in the cytogenetic map independently of their relations with other clones. We have termed a cytological map of such clones a "framework map".[8] Since virtually everything in *Drosophila* genetics is also referred to the polytene chromosome map, the polytene chromosomes serve as a bridge between the physical map and the linkage map. It is because of the historical primacy of the polytene chromosomes that the framework maps of P1 and YAC clones have great utility even though clones that appear to be adjacent in the framework map need not necessarily overlap at the molecular level. It therefore becomes possible to build an integrated genome map step by step even as assembly of the clones into molecular contigs is still being completed.

Table 5.1. Bands and DNA content*

Chromosome	Number of bands	Estimated DNA content (kb)
X	983	22,770
2L	874	18,301
2R	1063	22,750
3L	1034	21,180
3R	1200	24,130
4	40	1,290
Total	5194	110,421

*Data from Heino TI, Saura AO, Sorsa V. *Drosophila* Information Service 1994; 73:619-738.

RELIABILITY OF IN SITU LOCALIZATIONS

Caveat emptor is the watchword: anyone using the material should confirm the cytogenetic localization before investing any time or effort in further analysis. In spite of the fact that the cytogenetic localizations of the P1 and YAC clones are very useful, the accuracy of the assigned limits of the hybridization should not be overinterpreted. We are not saying that the localizations are wildly inaccurate. Data discussed below indicate that over 90% of the localizations are highly reproducible. However, about 3% of the clones may be discordant from the assigned localizations for any of a number of reasons, and so reasonable care should be taken to identify these clones in advance.

Limitations of the Procedures

There are at least four inherent limitations in the method of in situ hybridization that mandate caution. One limitation is that some regions of the genome are much more favorable for analysis than others. Among the most difficult regions are the bases of the chromosomes, virtually all of chromosome 4, and numbered sections 13-14 on the X and 35-37 on 2L. It seems that everyone who has analyzed polytene chromosomes has one or more favorite bugaboo regions. A second limitation is that the quality of the chromosome squashes and the in situ preparation is variable from one slide to the next. With practice and experience, a researcher can usually make preparations that are high enough in quality that they will contain nuclei with virtually any region optimally stretched to enable the limits of hybridization in that region to be determined with reasonable accuracy. Inevitably, some preparations are encountered that one wishes were better. If there were only one clone or a small number of clones to be localized, it would be possible to make multiple slides and choose the best nuclei for analysis. It is through multiple preparations that inversion breakpoints and the positions of single probes are localized with maximal accuracy. However, in a project in which thousands of clones need to be localized, one must inevitably offset the needs of accuracy against those of throughput.

A third limitation on accuracy is the inherent limit of resolution of the in situ hybridization signal. We have generally quoted the approximate limit of cytological resolution as 50 kb.[20] The use of biotinylated probes increases the resolution to perhaps 20 kb.[21]

Still finer resolution to 5-10 kb has been reported,[10] but this is possible only in the most favorable regions of the polytene chromosomes. Given that the average band contains approximately 20 kb, and the range is 5 kb to 90 kb (Fig. 5.1), it is clear that resolution to the individual band is often not possible. Yet the localization data in the appendices are tabulated with a discrete band at each end. These are "best shot" reads—the most accurate an experienced reader can determine, given the quality of the preparation. Users of the clones should understand that the true extent of the hybridization may sometimes be a few bands to either side. We opted for best-shot localizations in part to maximize throughput and in part in recognition of the fact that assigning a range to the hybridization at each end of a clone is also somewhat subjective and perhaps errs equally in suggesting greater accuracy than is warranted.

A final limitation that needs to be mentioned is that of possible errors in data handling or in the disposition of clones in the libraries. Data handling errors are mainly writing in the notebooks (or entering into the database) one letter or number when another was intended. For example, Cambridge clones 41-89 (DS08489) and 41-91 (DS08681) were originally entered into the database incorrectly as hybridizing with 35F1,2 and 88B1,2, respectively; the fact is the reverse—DS08489 hybridizes with 88B1,2 and DS08681 with 35F1,2. The mistake resulted from the fact that the localizations of the two clones were copied off the wrong (adjacent) lines in the notebook. Likewise, Cambridge clone 90-77 (DS07385) was originally entered into the notebook as hybridizing with 2A11;2A15. This assignment raised some eyebrows because these bands do not exist; the correct localization is 2B11;2B15, and the person analyzing the slide simply wrote "A" where "B" was intended. We believe that errors of this type in the data are very infrequent. Furthermore, virtually all of the errors will eventually be detected as inconsistencies in the course of the STS mapping, which is how the bookkeeping errors for DS08489 and DS08681 were found.

Inconsistent localizations may also originate from the restreaking and reorganization of the P1 libraries described in chapter 2. For example, if a few of the wells in the original Cambridge library contained two different clones, one of which was recovered for the original in situ localization and the other of which was selected for making the Berkeley library, then the localizations will disagree. Another possible source of inconsistency is misplacement of a clone from its position in the Cambridge library to its targeted position in the Berkeley library. These considerations are germane to the reported P1 localizations because approximately two-thirds of the in situ hybridizations were carried out with clones from the Cambridge library. Fortunately, there seem to be very few errors of this type.

REPEATABILITY OF P1 LOCALIZATIONS

As a form of quality control on the accuracy of the procedures, a total of 214 clones were deliberate duplicates introduced into the workstream as blind controls. These controls were designed primarily to check the correspondence between the Cambridge and Berkeley libraries. All of the blind controls were isolated from the Berkeley library and the localizations compared with those obtained previously from isolates of the supposedly corresponding clone from the Cambridge library. However, analysis of the blind controls also serves to ascertain the reproducibility of the cytological localizations.

The data from the duplicate preparations are summarized in Table 5.2. An identical assignment means that the two localizations agreed to the band for both ends of the clone. An overlapping assignment means that there was a small discrepancy in the localization at one or both ends. Considering only the 168 controls assigned to single euchromatic locations in both preparations, 91% of the assignments were either identical or overlapping. An-

Table 5.2. Repeatability of P1 localizations

Localization in salivary chromosomes	Number of clones
Identical assignments	78
Overlapping assignments	75
Small gap between assignments	10
Discordant assignments	5
Dual site versus single site	7
Dual sites in both preparations	4
Chromocenter versus single site	8
Chromocenter in both preparations	27
Total	214

other 6% of the clones were assigned to the same lettered subdivision but had a small gap between the localizations. Only 3% of the clones give completely discordant reads. Virtually all of the discordant assignments originate from clones containing moderately repetitive DNA in which the two preparations exhibited a different, but sufficiently prominent, site of hybridization to be identified as the major site of hybridization assigned to the clone. Although the number of completely discordant reads from this source is small, the fact that there are some emphasizes the caveat stated at the outset of this discussion: the in situ localization of any clone should be verified before investing additional time or effort. There may also be a few discrepancies resulting from such possibilities as data handling error, multiple clones in some wells of the Cambridge library, or lack of perfect correspondence between some clones in the Cambridge and Berkeley libraries.

Chromocentric Clones and Moderately Repetitive DNA

The controls in Table 5.2 also illustrate several issues regarding clones that hybridize primarily with the chromocenter or that contain moderately repetitive DNA. For clones hybridizing with the chromocenter, different preparations often give dissimilar degrees of labeling (heavy, moderate, or light), and so the annotation of the degree of labeling refers only to the particular preparation. However, certain unusual patterns of hybridization with the chromocenter that are observed with some clones, such as hybridization in discrete dots, do seem to be quite reproducible in different preparations.

Many P1 clones contain moderately repetitive DNA sequences that result in sites of secondary hybridization. In the analysis of these clones, the number of secondary sites that are clearly visible varies from one preparation to the next because of variation in the amount of label in the probe. As noted in connection with the five completely discordant assignments in Table 5.2, clones containing moderately repetitive DNA can also cause problems in interpretation because different sites may hybridize prominently enough to be assigned as the major site of hybridization in different preparations. Table 5.2 includes eight clones in which the primary site of hybridization was assigned to single euchromatic sites in one preparation but to the chromocenter in another. The clones with discordant assignments, as well as many of those with dual assignments (two euchromatic sites with primary hybridization) usually also result from different interpretations of the original and control preparations. The fact that some clones have discordant assignments resulting from the presence of transposable elements reinforces the necessity of confirming the assignment of any particular clone of potential interest.

REPEATABILITY OF YAC LOCALIZATIONS

The experience with repeatability of YAC localizations is not markedly different than found with P1 clones except that no extensive set of blind controls was carried out. However, the YAC libraries designated DY (random sheared inserts) and N (*Not*I fragments) contained many duplicate clones picked from the same colonies on the initial selective plates; as many as 25% of the DY clones and 10% of the N clones may be duplicates created in this manner.[10] Some of the same clones were also analyzed independently in both the Hartl laboratory and in the Duncan laboratory,[10,22] and many of the slides analyzed in the Duncan laboratory were read twice by different people.[10] Only one from each pair of duplicate clones is included in the tabulations.

The data on the repeatability of the analysis of duplicate YAC clones are quite comparable to those for P1 clones in Table 5.2. This result is not surprising for clones analyzed in the Hartl laboratory because the analysis of both P1 and YAC clones was carried out by the same people. As in the case of the P1 clones, the YAC assignments to euchromatic sites are usually repeatable to a close approximation, often to particular bands. Also similar to the P1 analysis, clones hybridizing with the chromocenter may yield a signal that is heavy, moderate, or light, depending on the preparation; and repetitive DNA elements present in the genome may or may not yield a visible signal depending on the degree of labeling of the probe.[10,22] In some cases secondary sites of hybridization due to repetitive DNA may be interpreted as major sites of hybridization. Therefore, the same degree of caution is necessary for dealing with YAC clones as for dealing with P1 clones.

COVERAGE ASSESSED IN TERMS OF INDIVIDUAL BANDS

Our first approach to estimating coverage of the euchromatic genome is to base the analysis on the fraction of the 5,198 bands that are included in one or more clones, irrespective of the variation in DNA content among bands.

P1 AND YAC CLONES

The distribution of the number of times each band is included in a P1 clone is illustrated in Figure 5.2. The distribution was obtained by tallying, for each P1 clone, each of the set of bands included within it, and then summing the tally marks for each band across the whole set of P1 clones. Because chromosome 4 is underrepresented in the P1 libraries (chapter 4), the tabulation in Figure 5.2 includes only chromosomes X, 2, and 3. Any literal interpretation of the distribution has to be hedged by the fact that the cytological assignments are not absolutely accurate or uniformly repeatable (Table 5.2). Nevertheless, here and elsewhere in our analysis, we assume that the cytological assignments are accurate

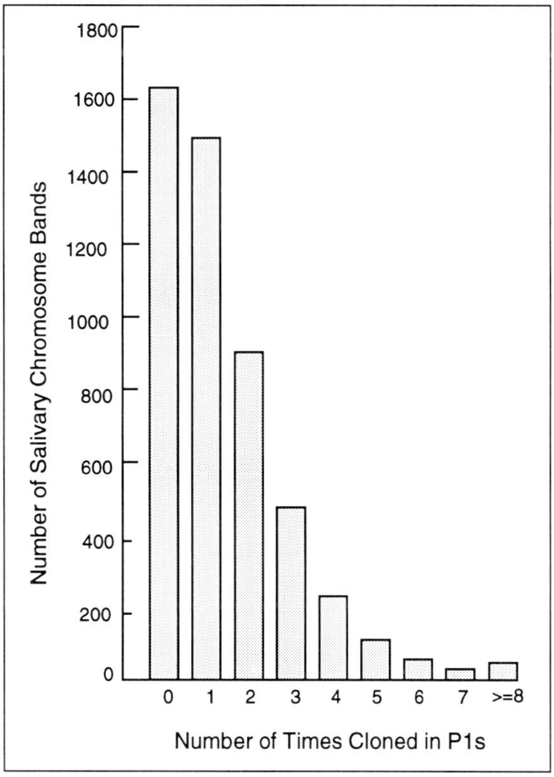

Fig. 5.2. Counts of polytene bands against redundancy of coverage (hits per band) in P1 clones.

enough on the whole that a genome-wide assessment of coverage is reliable as a first approximation. The large number of clones analyzed helps the situation because, in theory, the reassignment of a clone that opens up a gap in one place will often be offset by the reassignment of another clone that closes a gap somewhere else.

Among the total of 5,157 bands in Figure 5.2, 32.5% (1,678 bands) are not included in any P1 clone. Assuming Poisson coverage of bands in the euchromatic genome, the number of euchromatic genome equivalents in the set of clones equals -ln(0.325), or 1.12. However, the total number of clones in Figure 5.2 is 2,460 and, assuming an average DNA content of 80 kb per clone, the number of genome equivalents equals 197 Mb/110 Mb = 1.79. The discrepancy in the estimates is quite large inasmuch as the Poisson expectation of the proportion of uncloned bands from the latter estimate is 16.7%. Part of the reason for the discrepancy is that many of the apparently uncloned bands have an estimated size smaller than 10 kb and so would usually not be assigned as the band at the end of a region of hybridization. Another part of the discrepancy results from the fact that the X chromosome is underrepresented in the libraries by 25% relative to the autosomes (chapter 4). However, confining the analysis to the autosomes does not change the situation appreciably: 1.21 genome equivalents based on the number of uncloned bands versus 1.98 genome equivalents based on the number of P1 clones. An apparent excess of uncloned bands is also evident in the YAC data of Cai et al[10] in which 1.30 genome equivalents can be estimated based on the number of uncloned bands but 1.64 genome equivalents based on the amount of DNA in the clones.

Table 5.3. Estimated coverage of euchromatin in genome equivalents (GE)

Method of estimation	P1clones	YACclones	P1&YAC
GE = -ln(proportion of bands not hit)	1.123	2.210	2.976
GE = mean number of hits per band	1.507	2.359	3.868
GE = -ln(proportion of euchromatin in gaps)	1.320	2.397	3.324
GE = DNA content of clones*	1.794	2.245	3.963

*Calculated as 2467 P1 clones averaging 80 kb and1193 YAC clones averaging 207 kb; the haploid euchromatic genome size is taken as 110 Mb.

The coverage of the genome based on the proportion of uncloned bands is consistently the smallest of several independent estimates. The data for the P1 libraries, the YAC library, and the P1 and YAC libraries combined are summarized in Table 5.3. In each case, larger estimates of coverage result from analysis of the mean number of "hits" (times cloned) per band and from the estimated amount of DNA in the uncloned bands than from analysis of the number of uncloned bands.

For the YAC library, the distribution of hits per band for the major autosomes is shown in Figure 5.3. Assuming 2.245 as the number of genome equivalents in this set of clones (Table 5.3), the number of bands in each of the categories from 0 through 4 is within about 10% of the Poisson expectation. However, in the tail of the distribution there are relatively too many bands cloned five or more times.

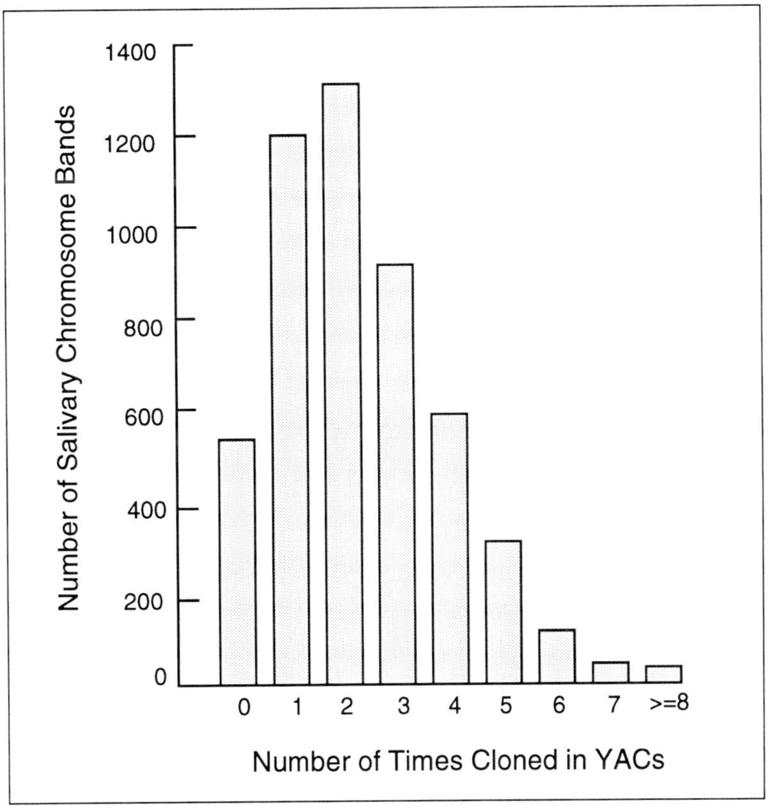

Fig. 5.3. Counts of polytene bands against redundancy of coverage (hits per band) in YAC clones.

Size Distribution of Mapped YACs

Figure 5.3 is the distribution of number of YACs per band. Because the mapped YAC clones range in size from less than 100 kb to more than 300 kb, it is of interest to turn the analysis on its head and ask, What is the distribution of number of bands per YAC? This distribution is illustrated in Figure 5.4. Because the average size of the mapped YACs is 206.8 kb, with a standard deviation of ±55.5 kb, and because the average band contains approximately 21 kb, the number of bands in the YAC clones is expected to have a mean of 9.8 and a standard deviation of ±2.6. The distribution in Figure 5.4 has a mean of 10.2 bands and a standard deviation of ±6.7 bands. The mean is in very good agreement with the naive expectation but the variance of the distribution is somewhat greater than expected. The inflated variance in the number of bands per YAC may simply reflect the fact that the bands are not distributed uniformly across the genome with respect to their DNA content. As noted in connection with Figure 5.1, the DNA content per band varies substantially among bands, ranging from a low of less than 5 kb to a high of 90 kb.[19] In the genome, some regions contain a higher than expected density of thick bands, other regions contain an apparent excess of very fine

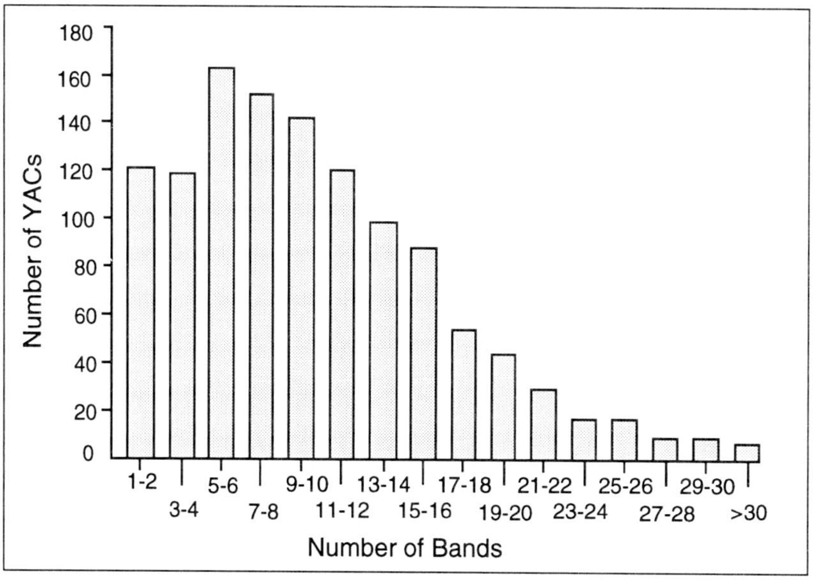

Fig. 5.4. Distribution of sizes of mapped YACs according to number of polytene bands included, as estimated from in situ hybridization.

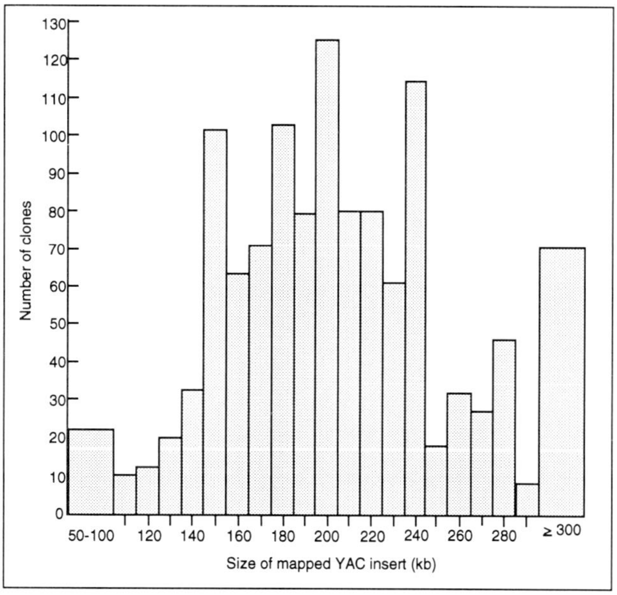

Fig. 5.5. Distribution of sizes of mapped YACs as estimated from pulse field gel electrophoresis (PFGE).

bands. One consequence of this phenomenon is that YACs in regions of thick bands will contain fewer bands than expected and YACs in finely banded regions will contain more. In our experience, YACs derived from regions of the genome where the staining intensity of the bands is light do tend to hybridize with many bands, while those derived from regions where the staining intensity is heavy tend to hybridize with few.[22] One consequence of a nonrandom distribution of bands across the genome is that the variance in the distribution of bands per YAC will be inflated relative to that expected from a random distribution.

The DNA content of a YAC can be estimated physically by pulse field gel electrophoresis (PFGE). The distribution of sizes of mapped clones is shown in Figure 5.5; this is a composite distribution consisting of 272 "DY" clones (random ends), 501 "N" clones (*Not*I fragments), and various numbers of clones containing *Eco*RI fragments—71 "E" clones, 305 "R" clones, 30 "Rt" clones, and 14 "RX" clones (total 1193). Chapter 2 contains detailed descriptions of the different categories of clones and the *Drosophila* strains that they are derived from. On the average, each YAC contains roughly the same quantity of DNA as the average lettered subdivision in the salivary gland chromosomes.[20]

An independent estimate of the sizes of the YACs can be determined by summing the estimated DNA content for each of the bands that the YAC contains. The relation between these estimates and those obtained from PFGE is shown in Figure 5.6, in which the dashed line depicts equal values along the ordinate and abscissa. The data are for the 196 clones assigned to 2L but the result is typical of all the major chromosome arms. As estimated from PFGE, the distribution of clone size is bunched around 200 kb but some clones are larger than 300 kb. Because the estimates are based on physical measurements, the PFGE values are certainly the more reliable. Estimates of clone size based on cytology are much more variable. The standard deviation among sizes, estimated from PFGE, is approximately 10% of the mean; that estimated from the cytology is approximately 50% of the mean. The cytological estimates include two sources of uncertainty. First, the bands included in any YAC are not always specified with complete accuracy (Table 5.2). Secondly, the estimates of DNA content per band are estimated indirectly from measurements of band width, band density, and other physical characteristics of the bands.

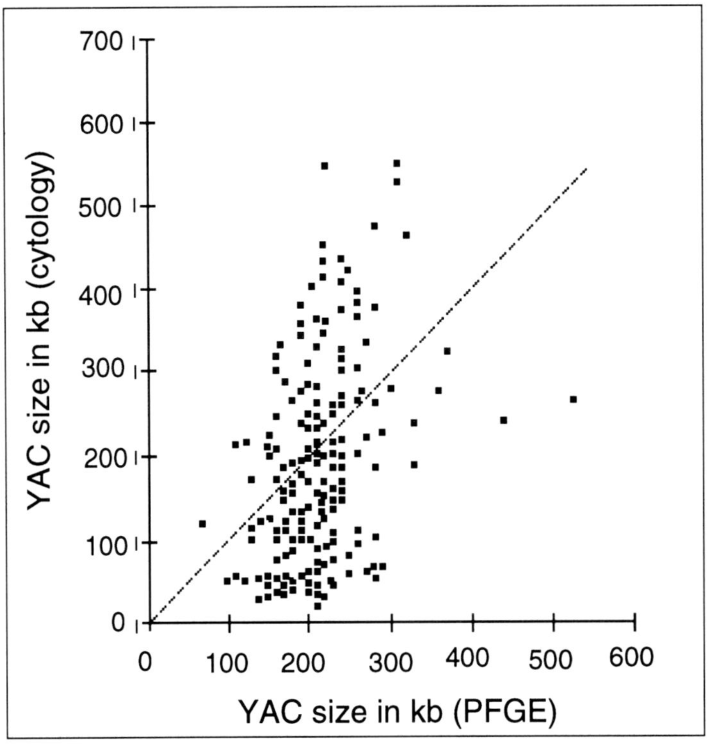

Fig. 5.6. Sizes of YACs assigned to 2L as estimated from pulse field gel electrophoresis (PFGE) and from estimated DNA content of the included polytene bands.

Given these uncertainties, it is not surprising that the cytological estimates of YAC content vary substantially more than the PFGE estimates. There is nevertheless a significant correlation ($r = 0.36$, $P < 0.01$).

It has been suggested that the cytological assignments of YAC clones are systematically biased toward excessive length because the hybridization signals extend artifactually beyond the actual limits of the clones.[10] In spite of the considerable sampling variance among the cytological estimates in Figure 5.6, the data do not support the suggestion of a systematic upward bias. If there were such a bias, a significant excess of the points in Figure 5.6 would fall above the line. There are, in fact, more points below the line, and the resulting bias is slightly but not significantly downward. Consequently, the mean size of the clones as estimated from PFGE is 213 kb and that estimated from cytology is 198 kb.

COMBINED P1 AND YAC COVERAGE

The combined P1 and YAC framework maps give very broad and deep coverage of most regions of the euchromatic genome. The smallest estimate of coverage in Table 5.3 (2.976) yields a Poisson coverage of 95% and the largest estimate (3.963) yields a

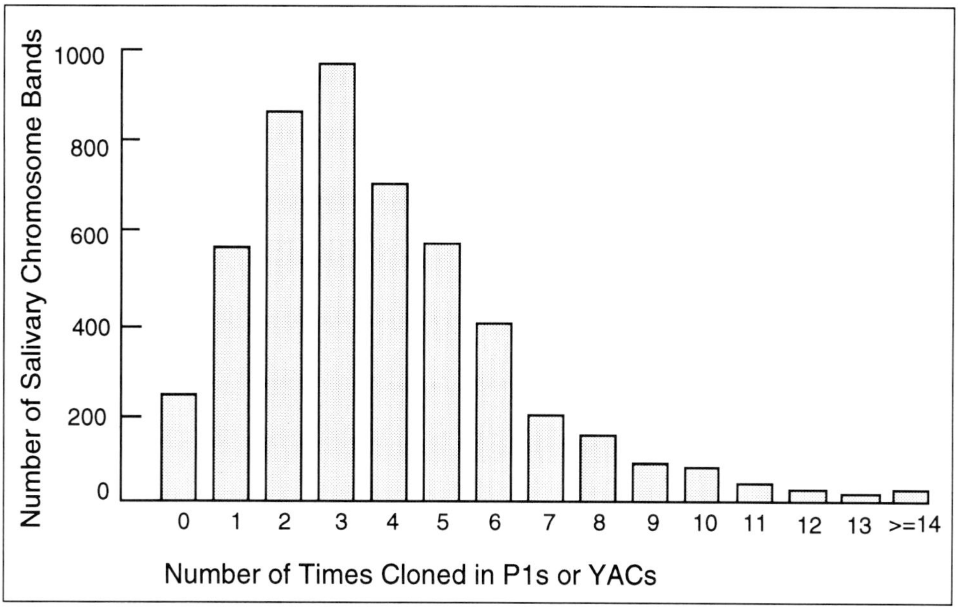

Fig. 5.7. Counts of polytene bands against redundancy of coverage (hits per band) in the mapped P1 and YAC clones together.

Poisson coverage of 98%. There are some gaps in coverage resulting from chance alone, and there may be other gaps because the DNA contained in them is difficult to clone. Extensive coverage of the euchromatin is also indicated by the data in Figure 5.7 for the distribution of hits per band in the P1 and YAC maps combined. Only 263 bands (5.1%) are not included in any clone and 4307 bands (83.5%) are included in two or more clones. The average number of hits per band in the combined maps is 3.87 (Table 5.3).

COVERAGE ASSESSED IN TERMS OF GAPS

The analysis in the previous sections treated each band as an independent unit and focused on the number of times each band had been cloned (hits per band). Generally speaking, the gaps in coverage are not isolated single bands but groups of contiguous bands. Analysis of such gaps is carried out in this section. Any such analysis is subject to the reliability and repeatability of the clone assignments as determined from in situ hybridization. Although the assignments are generally reliable as to region (Table 5.2), they are not infallible as to the particular bands. How-

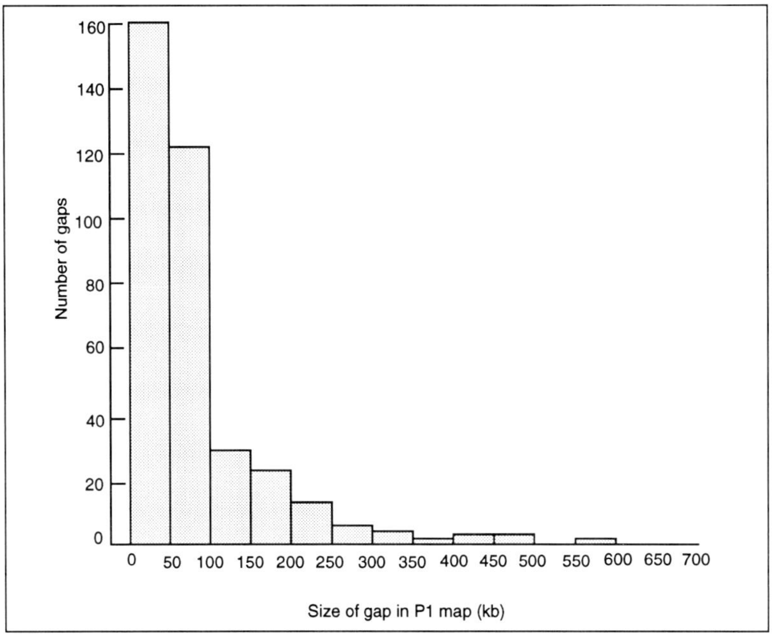

Fig. 5.8. Distribution of gap size in the framework P1 map. The size of each gap was determined by summing the estimated amount of DNA present in each band in the gap.

ever, slight misassignments are of less concern in a global analysis of large number of clones because of offsetting errors to be expected in different clones in the framework map. The size of the gaps in the framework map can be estimated either in terms of the number of contiguous bands that are not included among the mapped clones[10] or in terms of the estimated DNA content of the bands in the gap.[9] We use the estimated DNA content because it gives different weight to gaps consisting of many fine bands and gaps consisting of an equal number of heavy bands. Although there may be considerable sampling error in the estimate of DNA content of any particular band (Fig. 5.6), the DNA content summed across several bands is expected to be more reliable because of the law of large numbers.

P1 COVERAGE

The distribution of gaps in the P1 framework map is illustrated in Figure 5.8. Gap sizes were estimated by taking the cytological assignments at face value and summing, for each gap, the estimated amount of DNA in each band included in the gap. The framework P1 map includes 369 gaps, of which 44% are smaller than 50 kb and 33% are in the range 50 kb to 100 kb. That 77% of the gaps are of an estimated size smaller than 100 kb is of some significance because it implies that most of the gaps in the map can probably be covered by additional single P1 clones. Therefore, the current work of screening the P1 library with STS markers derived from sequences at the ends of P1 clones (chapter 6) is expected to close many of the gaps relatively quickly. On the other hand, there are 24 gaps with estimated sizes greater than 200 kb that will probably be more difficult to close.

Elementary probability considerations imply that, with n randomly chosen clones of fixed size d covering a genome of size L, the probability that a gap will be of size greater than x equals approximately $\exp[-n(x + d)/L]$. Hence, the proportion of gaps smaller than x is expected to be $1-\exp[-n(x + d)/L]$. Reasonable values for the P1 clones are $d = 80$ kb in a euchromatic genome of size $L = 110$ Mb with $n = 2467$ clones mapped. With these values, one expects 98% of the gaps to be smaller than 100 kb and only 0.2% of the gaps to be greater than 200 kb. These estimates are way off: the observed values are 77% gaps smaller than 100 kb and 6.5% larger than 200 kb. Relative to the theory, the observed data have too few short gaps and too many long ones. The observed percentages for gap sizes < 100 kb and > 200 kb fit a theoretical model with a number of clones containing an amount of DNA in the range 0.65 to 0.80 genome equivalents. Evidently, some regions of the euchromatic genome are more difficult to clone than others, resulting in the excess of large gaps. This type of distribution is also seen in the mapped YAC clones.

YAC Coverage

The estimated sizes of the gaps in the framework YAC map are shown in Figure 5.9. There are 92 gaps altogether, of which 38% are smaller than 50 kb and 29% are between 50 kb and 100 kb. Nineteen of the gaps are greater than 150 kb. The distribution of gaps in Figure 5.9 differs from than reported previously[9] because, in the tabulation for Figure 5.9, the assignments of the YAC clones were taken literally; the main difference from the previous analysis is that, in the previous analysis, each YAC assignment was arbitrarily shortened by one band on each end in the interest of being conservative. For comparison, the previous analysis yielded 154 gaps of average size 90 kb. The reason for being

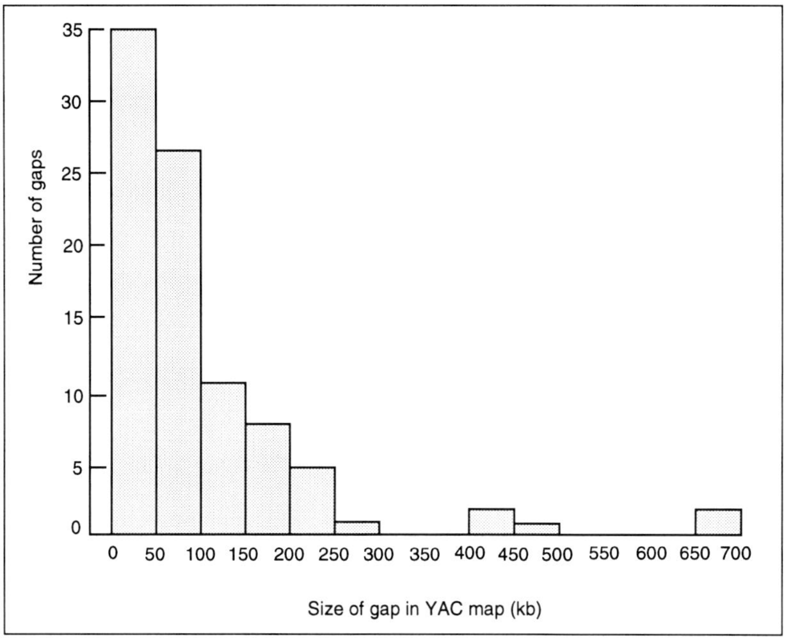

Fig. 5.9. Distribution of gap size in the framework YAC map.

less conservative in the present instance is to be able to use the same assumptions in the analysis of the YAC clones as in the analysis of the P1 clones so that relative coverage of the framework maps can be assessed on the same basis.

As observed with the P1 clones, there is a relative deficiency of small gaps and an excess of large gaps compared with the theoretical expectation. Taking the values $d = 200$ kb for the size of each YAC (assumed, for the sake of simplicity, to be identical for all YACs) and $L = 110$ Mb as before, then with 1193 mapped clones one expects 96% of the gaps to be smaller than 100 kb and only 1.3% of the gaps to exceed 200 kb. A theoretical model in which coverage is based on a number of clones equaling only one genome equivalent yields predictions close to the observed values of 77% for gaps smaller than 100 kb and 21% larger than 150 kb. Again, the discrepancy from theoretical expectation suggests that some regions of the euchromatic genome are underrepresented in the clones.

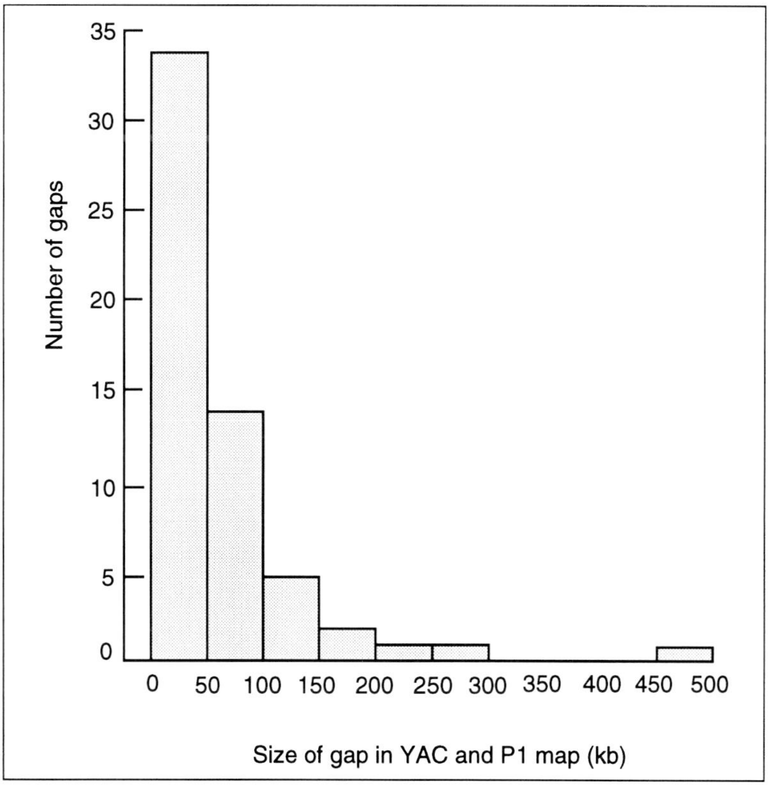

Fig. 5.10. Distribution of gap size in the combined P1 and YAC framework maps.

Locations of the Largest Gaps

The cytological locations of the largest gaps in the framework maps are summarized in Table 5.4. The P1 data includes all gaps of estimated size greater than 200 kb; the YAC data includes all gaps of estimated size greater than 150 kb. The combined P1 and YAC data includes gaps greater than 100 kb that are apparently not covered in either of the framework maps. This list identifies regions either that were very unlucky in being excluded from both mapping projects or that are unaccountably difficult to clone and underrepresented in the libraries. The distribution of estimated gap sizes in the combined P1 and YAC maps is illustrated in Figure 5.10.

Among the largest gaps in the combined P1 and YAC data are gaps including all or part of 6CD, 16BC, 18DE, 31E, and 81F.

Table 5.4. Locations of largest cytological gaps

P1 Gap	Size (kb)	YAC Gap	Size (kb)	YAC & P1 Gap	Size (kb)
(>200 kb)		(>150 kb)		(>100 kb)	
1A1;1A8	218	2C8;2D6	243	2D2;2D6	136
1B10;1C3	233	6C3;6E7	665	6C3;6D8	495
2D2;2E3	223	7D10;7D22	248	11B10,11;11B17	146
6C3;6D8	495	11B4;11B17	223	15B4;15C6	107
7E7,8;7F10	262	14D1,2;14E4	165	16B6;16C10	277
10C3;10D2	238	16A5,6;16E4,5	685	18D5;18E5	228
12B3;12C7,8	291	18C5;18E5	403	29F4;29F8	107
13C3;13E11	369	21A1;21B1	199	31E2;31E7	165
15A6;15F4,5	587	29E5;29F8	229	35F6;35F12	102
16B5;16C10	277	31E2;32A1	404	81F1;81F6	189
18D5;18E5	228	32C3;32E2	292		
25A5;25C4	481	35F6;36A2	175		
42C3;42E4	406	37E4;37F5	180		
47A4;47A16	271	43B1;43C4	237		
51F3;52A9	300	60A3;60A11	159		
55C6;55D4	232	68F1;68F8	195		
63B9;63C6	239	81F1;81F6	189		
64F4;65A5	210	85F3;86B1,2	485		
69A1;69C1	312	89E10;89F4	175		
82F10;83A9	200				
90B3;90C6	233				
93B7;93C7	228				
93F11;94A16	402				
95F2;95F15	228				

The gaps in 6CD, 16BC, and 31E persist even with a somewhat larger set of YACs including many chimeric clones not included in the present tabulation.[10] It is of interest that six of the ten largest gaps in Table 5.4 are on the X chromosome. This concentration probably reflects that fact that the X chromosome is generally more difficult to cover than the autosomes because of the theoretical 3 : 4 ratio of X clones : autosomal clones in the libraries; in the P1 libraries, the excess of autosomal clones is even greater.

The largest "gap" not included in Table 5.4 consists of chromosome 4, to which only 7 P1 clones and 3 YAC clones are assigned. More than 50% of chromosome 4 is not included in either the P1 or the YAC framework maps, and the largest single gap is 438 kb from 102D3 to 102F4. The underrepresentation of chromosome 4 is also apparent in other YAC mapping data.[10] Why this chromosome is so severely underrepresented in the libraries is unclear. One plausible hypothesis is that the DNA composition of chromosome 4 is responsible because chromosome 4 consists largely of beta heterochromatin and contains an abundance of repetitive DNA.

Table 5.5. Gaps by chromosome arm

Arm		P1 map	YAC map	P1 and YAC map
X	Number of gaps	73	24	18
	Mean (kb)	114	147	109
	Standard deviation (kb)	106	189	121
2L	Number of gaps	65	16	12
	Mean (kb)	65	124	59
	Standard deviation (kb)	67	113	47
2R	Number of gaps	76	14	8
	Mean (kb)	78	70	32
	Standard deviation (kb)	70	65	24
3L	Number of gaps	66	19	11
	Mean (kb)	67	74	37
	Standard deviation (kb)	63	46	18
3R	Number of gaps	89	18	9
	Mean (kb)	68	99	59
	Standard deviation (kb)	64	109	57

Distribution of Gaps by Chromosome

The gaps in coverage of the euchromatin in the major chromosome arms are summarized in Table 5.5. These are the largest, most conspicuous gaps. The framework maps undoubtedly also contain additional, generally smaller gaps between clones that terminate in the same band but do not overlap at the molecular level or that result from minor errors in the assessment of the cytological extent of hybridization. Taking the data in Table 5.5 at face value, the P1 framework map has 369 gaps averaging 79 kb, the YAC map has 91 gaps averaging 106 kb, and the combined P1 and YAC map has 58 gaps averaging 67 kb. Roughly speaking, the standard deviation of the distribution of gaps is approximately equal to the mean. Gaps in the X chromosome are more numerous and, on the average, larger than those in the autosomes, undoubtedly reflecting the expected 25% deficiency of X-chromosomal clones, relative to autosomal clones, in the libraries. Considering only the autosomal gaps in Table 5.5, the average sizes are 70 kb, 92 kb, and 48 kb for the P1, YAC, and combined framework maps, respectively.

OVERALL ASSESSMENT OF COVERAGE OF THE EUCHROMATIC GENOME

The P1 and YAC maps of *Drosophila* were the first framework maps of any organism based on systems designed for cloning large DNA molecules. When the YAC mapping began, there were legitimate concerns about the possible instability of YAC clones, the possibility that many genomic sequences might be unclonable or inefficiently cloned in YACs, potential difficulties in screening YAC libraries, the occurrence of chimeric YACs with DNA sequences from noncontiguous genomic regions, and the general inexperience of most laboratories in working with yeast cells and with large DNA molecules. In the past few years many of these concerns have moderated.[23] Data from humans and a variety of model organisms, including *Drosophila*, show that most YAC clones are stable, their inserts are colinear with the genomic sequences from which they derive, and there appears to be no systematic bias in the euchromatic sequences that can be cloned. Screening protocols using colony hybridization have been improved, and methods based on the polymerase chain reaction have been developed that are particularly efficient for screening large libraries.[24]

There is even a first-generation map of the human genome.[25] Experience in working with large DNA molecules has spread in the scientific community, and strategies and protocols for working with YACs are becoming widely available.[23] On the other hand, in spite of the technical progress, YACs are still not as easy to work with as bacterial host-vector systems. Furthermore, in some YAC libraries, the very substantial frequencies of chimeric clones, in which two DNA fragments from noncontiguous regions of the genome are joined together, remains a serious technical obstacle to efficient mapping. The frequency of chimeric clones in some human YAC libraries is 50% or more.[26] Apparently, many of the chimeric clones in human libraries result from recombination in the host between a centromere-bearing and a telomere-bearing fragment containing highly repetitive DNA sequences, such as the Alu repeat. The *Drosophila* genome contains fewer highly repeated sequences than the human genome, and the YAC libraries of *Drosophila* also have fewer chimeric clones; estimates of the frequency of chimeric *Drosophila* clones range from 5% to 10%.[10,22] In the present analysis of the YAC framework map, most of the apparently chimeric YAC clones have been excluded.

The P1 cloning system has also performed very well in its coverage of the euchromatic genome of *Drosophila*. Yields of DNA from P1 clones are very high relative to YACs.[7,27,28] The ends of genomic fragments inserted in P1 clones can be amplified by PCR[29] or their sequences determined by cycle sequencing. There is even a rapid and convenient method for systematically subcloning the inserts in P1 clones into overlapping fragments of any desired size for use in genomic sequencing and for other purposes.[30]

Assessed in terms of bands in the polytene salivary gland chromosomes, the framework P1 map includes 67% of the 5,157 bands in the large chromosomes (excluding chromosome 4). The YAC map includes 89% of the bands, and the combined P1 and YAC maps include 95% of the bands. Coverage in terms of estimated DNA content is not markedly different: 73% for the P1 map, 91% for the YAC map, and 96% for the combined map. There is, however, a statistical excess of large gaps, which may indicate that some euchromatic regions are underrepresented in the P1 and YAC libraries. Clones from chromosome 4 are severely underrepresented to about the same extent in both the P1 and YAC libraries. Among the mapped P1 clones, 0.28% are assigned to chromosome 4; and among the mapped YAC clones, 0.25% are assigned to chromosome 4. Approximately 1% of chromosome 4 clones would be expected, and so this chromosome is underrepresented by approximately a factor of 4. Other regions of the euchromatin may also be underrepresented, although perhaps not so severely, which would account for some of the larger than expected gaps.

Coverage of the heterochromatin is still quite uncertain because the positions of clones hybridizing with the chromocenter of salivary gland nuclei cannot be resolved. Greater resolution in this part of the genome will require in situ hybridization with mitotic chromosomes to assign the clones to chromosome arms and to regions within arms. Approximately 12.5% of the P1 clones hybridize primarily with the chromocenter but the P1 libraries contain few, if any, clones composed primarily of simple satellite sequences (chapter 4). The proportion of chromocentric clones is approximately the same as the estimated amount of heterochromatin not including simple satellite sequences. Therefore, it is possible that the heterochromatic P1 clones contain a substantial proportion of the sequences that are genetically important, including coding sequences, repeated blocks of transposable elements, and so on. Some of the results of screening the P1 libraries with known heterochromatic sequences of these types are discussed in chapter 4.

References

1. Sturtevant AH. The linear arrangement of six sex-linked genes in *Drosophila*, as shown by their mode of association. J Exper Zool 1913; 14:43-59.
2. Bouche JP, Gelugne JP, Louarn J, Louarn JM, Kaiser K. Relationships between the physical and genetic maps of a 470 x 10^3 base-pair region around the terminus of *Escherichia coli* K12 DNA replication. J Mol Biol 1982; 154:21-32.
3. Olson MV, Dutchik JE, Graham MY, Brodeur GM, Helms C, Frank M, MacCollin M, Scheinman R, Frank T. Random-clone strategy for genomic restriction mapping in yeast. Proc Natl Acad Sci USA 1986; 83:7826-7830.
4. Coulson A, Sulston J, Brenner S, Karn J. Toward a physical map of the genome of the nematode *Caenorhabditis elegans*. Proc Natl Acad Sci USA 1986; 83:7821-7825.
5. Burke DT, Carle GF, Olson MV. Cloning of large segments of exogenous DNA into yeast by means of artificial chromosome vectors. Science 1987; 236:806-812.

6. Sternberg NL. Cloning high molecular weight DNA fragments by the bacteriophage P1 system. Trends Genet 1992; 8:11-16.
7. Pierce JC, Sternberg NL. Using the bacteriophage P1 system to clone high molecular weight genomic DNA. Methods Enzymol 1992; 216:549-574.
8. Hartl DL, Nurminsky DI, Jones RW, Lozovskaya ER. Genome structure and evolution in *Drosophila:* Applications of the framework P1 map. Proc Natl Acad Sci USA 1994; 91:6824-6829.
9. Hartl DL. Genome map of *Drosophila melanogaster* based on yeast artificial chromosomes. In: K. E. Davies, S. M. Tilghman, eds. *Genome Analysis Volume 4: Strategies for Physical Mapping.* Cold Spring Harbor, NY: Cold Spring Harbor Laboratory Press 1992; 39-69.
10. Cai H, Kiefel P, Yee J, Duncan I. A yeast artificial chromosome clone map of the *Drosophila* genome. Genetics 1994; 136: 1385-1401.
11. Rudkin GT. Cytochemistry in the ultraviolet. Microchem J Symp Ser 1961; 1:261-276.
12. Ashburner M. *Drosophila: A Laboratory Handbook.* Cold Spring Harbor, NY: Cold Spring Harbor Laboratory Press 1989.
13. Bridges CB. Salivary chromosome maps with a key to the banding of the chromosomes of *Drosophila melanogaster.* J Heredity 1935; 26:60-64.
14. Bridges CB. A revised map of the salivary gland X-chromosome of *Drosophila melanogaster.* J Heredity 1938; 29:11-13.
15. Bridges PN. A new map of the salivary gland 2L-chromosome of *Drosophila melanogaster.* J Heredity 1942; 33:403-408.
16. Bridges CB, Bridges PN. A new map of the second chromosome: a revised map of the right limb of the second chromosome of *Drosophila melanogaster.* J Heredity 1939; 30:475-476.
17. Bridges PN. A revised map of the left limb of the third chromosome of *Drosophila melanogaster.* J Heredity 1941; 32:64-65.
18. Bridges PN. A revision of the salivary gland 3R-chromosome map of *Drosophila melanogaster.* J Heredity 1941; 32:299-300.
19. Heino TI, Saura AO, Sorsa V. Maps of the salivary gland chromosomes of *Drosophila melanogaster. Drosophila* Information Service 1994; 73:619-738.19.
20. Sorsa V. *Chromosome Maps of Drosophila.* 2 Volumes. Boca Raton, Florida: CRC Press 1988.
21. Merriam J, Ashburner M, Hartl DL, Kafatos FC. Toward cloning and mapping the genome of *Drosophila.* Science 1991; 254:221-225.
22. Ajioka JW, Smoller DA, Jones RW, Carulli JP, Vellek AEC, Garza D, Link AJ, Duncan IW, Hartl DL. *Drosophila* genome project: One-hit coverage in yeast artificial chromosomes. Chromosoma 1991; 100:495-509.
23. Brownstein BH, Nelson DL. *YAC Libraries: A User's Guide.* New York, W. H. Freeman 1994.

24. Green ED, Olson MV. Systematic screening of yeast artificial-chromosome libraries by use of the polymerase chain reaction. Proc Natl Acad Sci USA 1990; 87:1213-1217.
25. Cohen D, Chumakov I, Weissenbach J. A first-generation physical map of the human genome. Nature 1993; 366:698-701.
26. Green ED, Olson MV. Chromosomal region of the cystic fibrosis gene in yeast artificial chromosomes: A model for human genome mapping. Science 1990; 250:94-98.
27. Sternberg N. Bacteriophage P1 cloning system for the isolation, amplification, and recovery of DNA fragments as large as 100 kilobase pairs. Proc Natl Acad Sci USA 1990; 87:103-107.
28. Smoller DA, Petrov D, Hartl DL. Characterization of bacteriophage P1 library containing inserts of *Drosophila* DNA of 75-100 kilobase pairs. Chromosoma 1991; 100:487-494.
29. Nurminsky DI, Hartl DL. Amplification of the ends of DNA fragments cloned in bacteriophage P1. Biotechniques 1993; 15:201-208.
30. Yoshida K, Strathmann MP, Mayeda CA, Martin CH, Palazzolo MJ. A simple and efficient method for constructing high resolution physical maps. Nucleic Acids Res 1993; 21:3553-3562.

CHAPTER 6

Strategy for Contig Building

One goal of physical mapping is to generate a set of ordered, overlapping clones covering the target genome for use in subcloning, screening, genomic sequencing, and other purposes. Complete, continuous coverage of complex genomes or long chromosomes is rarely possible in practice because some small regions of the genome may be absent (and almost inevitably are absent) from even large, highly redundant libraries. Nevertheless, it is possible to obtain virtually complete coverage of the genome piecewise in the form of long, continuous, unbroken stretches of overlapping clones.[1-3] Such groups of overlapping clones are called "contigs." Many physical mapping strategies rely on a "bottom-up" strategy in which clones with relatively small inserts are sorted into contigs by fingerprinting, restriction mapping, or other methods.[1-5] This method is efficient but has the limitation that the physical map comes together quite late in the experimental program because long-range continuity requires that all segments of the target genome be cloned and analyzed. More rapid closure can be achieved by adding a "top-down" component to the strategy, which usually takes the form of a long-range restriction map[6] or clones containing long inserts (YACs or P1s) to connect adjacent, nonoverlapping contigs.[7]

In the *Drosophila* genome project, the top-down component of physical mapping gained primacy in order to take full advantage of the polytene salivary gland chromosomes and to localize large numbers of P1 and YAC clones for immediate distribution to the research community. This chapter presents the current coverage of the genome as assessed by apparently overlapping groups of clones in the framework map. The PCR-based screening strategy for identifying contigs at the molecular level is also discussed.

CYTOLOGICAL CONTIGS AND ISLANDS

In an earlier analysis of YACs, the clones were sorted into groups within which there were no obvious gaps in coverage of the genome as assessed by in situ hybridization with polytene chromosomes.[8] These groups were called "cytological contigs." In a practical sense the clones in cytological contigs are "contigs by default" since they occupy the space between obvious gaps.[9] In this book we will call such groups of clones "islands" but without any implication that the islands necessarily overlap at the molecular level.[10-12] That is, an "island" of clones is a group of clones mapping so close together in the framework map that their coverage of the genome is not interrupted by any obvious gaps. Like a real island, an island of clones may, in fact, be cut across by one or more small gaps that interrupt its physical continuity. Identification of islands depends on the relatively high cytological resolution of the polytene salivary gland chromosomes, but there are two principal inherent limitations in the method. First, while the cytological localizations of the clones are generally accurate, the specification of the particular bands that mark the end points is based on visual inspection and so inevitably has a degree of subjectivity depending to a significant extent on the quality of the banding in the relevant region and also upon the degree of labeling and the efficiency of hybridization. The identification of the ends of the clones is therefore not absolutely reproducible in duplicate preparations (Table 5.2 in chapter 5) or even in independent reads of the same preparations.[9] Secondly, the accuracy of the estimates of DNA content per band are uncertain. They are based on the thickness and density of the bands as observed in the electron microscope in thin sections,[13] and they may be in error to the extent that the bands are nonuniform in different planes of section or in different genomic regions. Hence, the analysis of islands should be regarded as approximate and having as its main purpose a summary of the *Drosophila* physical map at its present stage of development in terms that allow comparison with genome projects in other organisms. The ultimate proof of clone overlap is their sharing of DNA sequences, and methods for establishing molecular contigs are discussed later in this chapter.

P1 Islands

The P1 framework map includes 367 islands with mean size approximately 200 kb and median size approximately 150 kb. There are 77 islands consisting of single, isolated clones and, when these are excluded, the remaining islands have mean and median size of 270 kb and 200 kb, respectively.

Distribution of Island Size

The size distribution of the P1 islands is shown in Figure 6.1. For islands smaller than 1 Mb the scale is in kb, and for those larger than 1 Mb the scale is in Mb. The size of each island was estimated from the data of T. I. Heino, A. O. Saura and V. Sorsa[13] by summing the DNA content of each band included in the island.

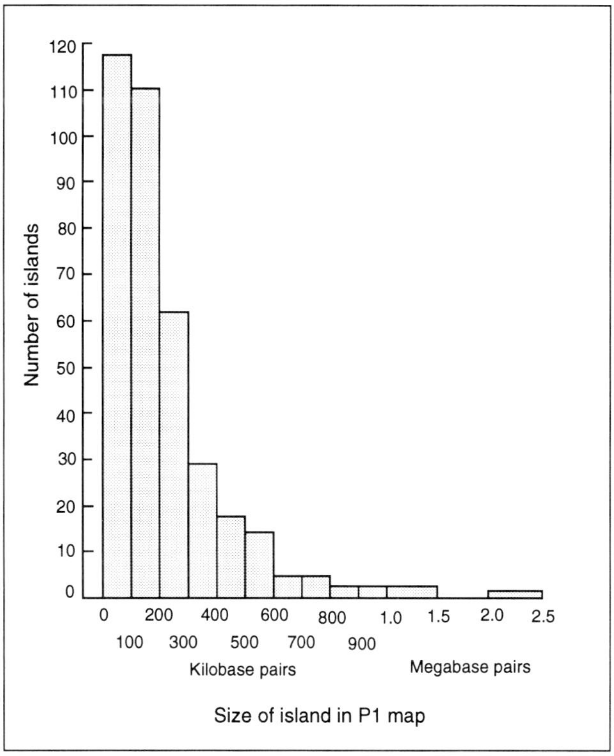

Fig. 6.1. Size distribution of P1 islands.

The DNA content of the interbands was not incorporated explicitly, since the estimated DNA content per interband averages less than 1 kb.[14] However, the DNA content per band estimated by Heino and collaborators[13] was obtained by normalizing to a total euchromatic genome size of 110 Mb, and so the interbands are implicitly taken into account.

The 367 islands in Figure 6.1 include an estimated 73.3% of sequences in the euchromatic genome as a whole. The three largest islands extend from 21A1; 21F3 (1.1 Mb), 23F1,2; 25A4 (1.1 Mb), and 3E6; 5D5 (2.5 Mb). We would caution again that the islands be interpreted in light of the limitations of cytological analysis discussed earlier: although there are no obvious gaps when the cytological assignments are taken at face value, there is no assurance that the islands represent a continuous group of clones.

Fig. 6.2. Distribution of contig size in a simulated library of 2467 clones, each 80 kb, covering a genome of 110 Mb.

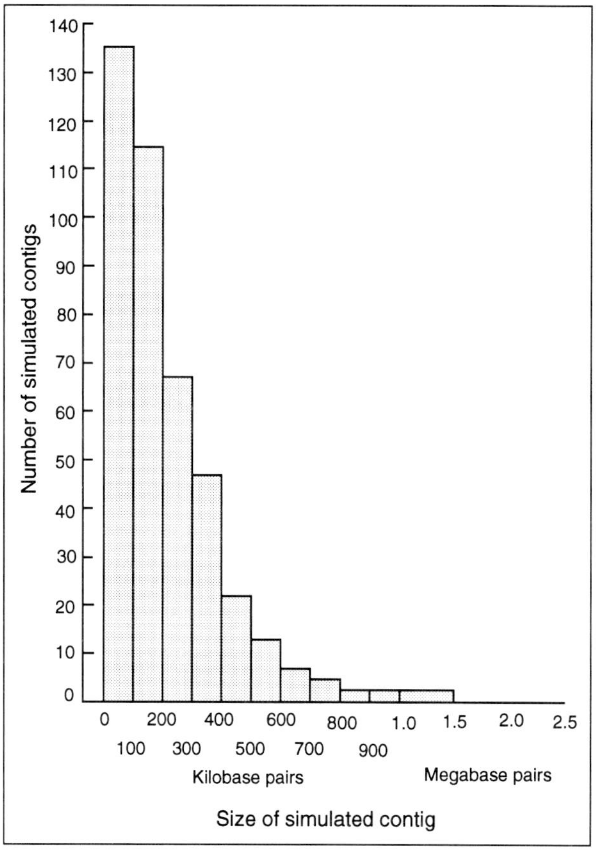

Nevertheless, it is clear that the locations of the large islands result from a dense clustering of clones in a particular region, and therefore these regions are fertile ground for true contig building.

The distribution of island size in Figure 6.1 is remarkably similar to the distribution of contig size in computer simulations of genome coverage matching the parameters of genome size, average clone size and number of clones. Figure 6.2 shows the distribution of contigs in a library consisting of 2467 clones with inserts of size 80 kb taken at random from a genome of size 110 Mb. The simulations were performed using a computer program written by Stanley A. Sawyer.[15] Figure 6.2 is a typical outcome. There are 426 contigs (compared with 367 P1 islands in Fig. 6.1) and there are a few more contigs in the range 0 to 100 kb and a few less in the range greater than 1.5 Mb (in fact, no contigs are larger than 1.2 Mb). Nevertheless, the overall similarity of the distributions in Figures 6.1 and 6.2 suggests that, while there are certainly gaps in some of the P1 islands, particularly in the largest ones, the islands may not be as gappy as intuition or good-natured skepticism may suggest.

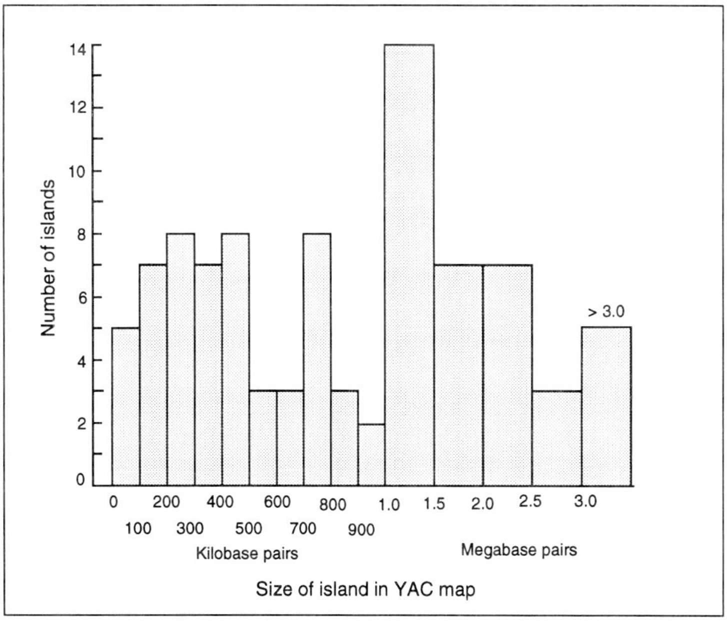

Fig. 6.3. Size distribution of YAC islands.

ISLANDS IN THE COMPOSITE FRAMEWORK MAP

Island size in the YAC framework map is distributed as shown in Figure 6.3. Altogether there are 90 islands averaging a little over 1 Mb; the median size is approximately 700 kb. As with the P1 islands, some of the YAC islands, particularly the largest ones, undoubtedly contain one or more small gaps interrupting the continuity.

Figures of conventional size are too small to illustrate the positions of the YACs on the chromosome map with an acceptable degree of resolution. However, the October 11, 1991, issue of Science (254: 165-344) includes a 55 cm × 80 cm fold-out poster with a drawing of the salivary gland chromosomes, which also shows the positions of the YACs along with cosmid contigs, lengthy chromosome walks, many genes, and much other useful information.[16] An update of the poster, including the locations of P1 clones accompanies this book.

Chimeric clones containing segments from two or more distinct regions of the genome have been common in some YAC libraries from the earliest stages of the technology.[17,18] The frequency of chimeric clones is particularly high in some human YAC libraries, perhaps because many of them result from recombination between repeated DNA elements, such as Alu repeats, in the target DNA. Chimeric YACs are less frequent in the *Drosophila* libraries,[9,19] and most of the recognized chimeras have been eliminated

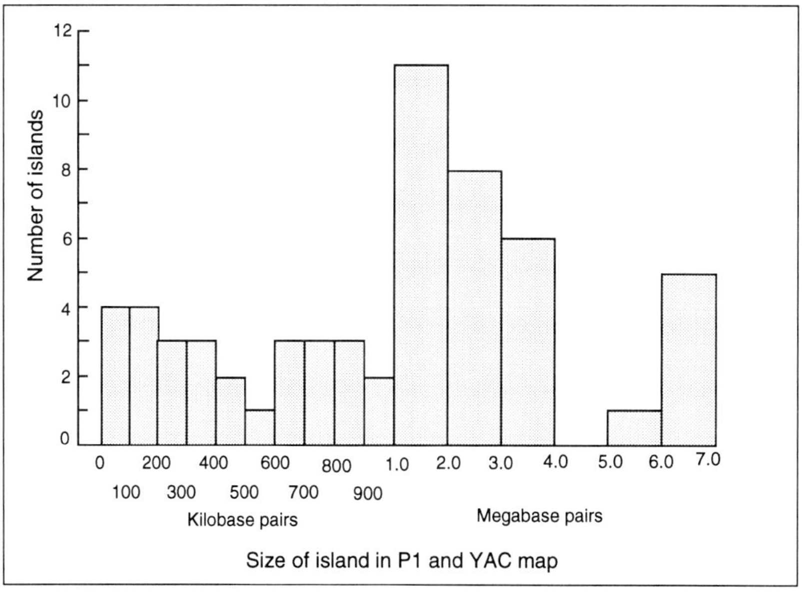

Fig. 6.4. Size distribution of islands in the composite P1-YAC map.

from the listing and analysis in this book. As far as we are aware, the YACs listed in this book include only six chimeras. They have been retained in the list because they were included in previous analyses[8,16] and only later found to be chimeric.[9] The list may also contain a few undetected chimeras. The known chimeras in the list are as follows: clone N13-61 assigned to 3B5-6; 3C1-2 also hybridizes with 67D; clone N09-07 assigned to 3E1; 3E2 also hybridizes with 67F; clone Rt05-24A assigned to 13A5-6 also hybridizes with 13BC and 22B; clone N23-77 assigned to 62B10-11; 62C2-3 also hybridizes with 42A; clone N24-27 assigned to 71B4-8; 71C1-2 also hybridizes with 84D; and clone N15-42 assigned to 82E1; 82E3-4 also hybridizes with 21C.

In the composite P1 and YAC map, there are 59 islands (Fig. 6.4). The average island size is approximately 1.8 Mb and the median size is approximately 1.0 Mb. Although all of the islands are not necessarily continuous at the molecular level, the islands nevertheless represent regions with an extensive accumulation of P1 and YAC clones and should certainly prove valuable in cloning genes in any of the regions.

The number and size distribution of islands in each of the major chromosome arms is shown in Table 6.1 for the P1, YAC,

Table 6.1. Islands by chromosome arm

Arm		P1 map	YAC map	P1 and YAC map
X	Number of islands	73	24	19
	Mean (kb)	199	802	1096
	Standard deviation (kb)	299	684	946
2L	Number of islands	65	16	12
	Mean (kb)	217	1017	1466
	Standard deviation (kb)	229	1232	1494
2R	Number of islands	77	13	9
	Mean (kb)	218	1558	2499
	Standard deviation (kb)	157	1160	2565
3L	Number of islands	66	19	11
	Mean (kb)	250	1039	1886
	Standard deviation (kb)	222	1284	2467
3R	Number of islands	87	18	8
	Mean (kb)	203	1239	2950
	Standard deviation (kb)	187	923	2670

and composite framework maps. The standard deviation of island size is approximately equal to the mean.

PHYSICAL MAPPING WITH SEQUENCE-TAGGED SITES

Identification of islands in the framework map, although a useful assessment of overall coverage, is inadequate for the selection of clones to be used as substrates in genomic sequencing because, for sequencing, continuity of coverage is essential.

The assembly of contigs of clones known to overlap at the molecular level can be carried out with any of a variety of experimental procedures such as restriction mapping,[1,2,6] fingerprinting,[3] hybridization with oligonucleotides,[4,20] or chromosome walking.[21] For clones with large inserts, a convenient strategy makes use of oligonucleotide primers to specifically amplify unique sequences in the target genome by means of the PCR.[17,22,23] Each such unique sequence is called a "sequence tagged site," or STS.[22] Any clone containing a particular STS can be identified by PCR because pools of DNA containing the clone will support amplification of the STS. Furthermore, any set of clones holding an STS in common must overlap, and so contigs can be identified and the clones within them ordered by means of the STS markers they contain.

Among the attractive features of STS content mapping is the usually unambiguous nature of PCR amplification, the virtually unlimited number of STS markers, and the availability of robots to carry out the repetitive operations of the assay. The efficiency of the method is strongly influenced by the choice of STS markers, however, and a number of alternative possibilities are discussed in the following sections.

MAPPING WITH RANDOMLY CHOSEN STS MARKERS

Strategies of STS content mapping initially envisioned a random choice of STS markers distributed throughout the genome.[22] One rationale for this strategy was to completely separate the set of STS markers from the set of mapping clones so that, as cloning methods improved and consecutive generations of vectors were developed, the physical map would not be wedded to a particular cloning technology. Although appealing in principle, the choice of random STS markers is inefficient in practice, and a number of alternatives have been proposed.[15,21]

The principal limitation in random STS selection is that the contig generated with each STS is relatively small. The small size results from the random position of the STS marker with respect to the cloned inserts that it is used to identify. For example, if a single, randomly chosen STS is used to screen a five-hit library of clones, the resulting contig is expected to include an average of five clones and extend, on the average, a total length of 1.6 times the size of the cloned insert.[15] A second limitation of random STS selection is that, when the number of STS markers becomes large, each successive STS has a diminishing effect in increasing contig size and decreasing contig number. At a time when a large number of randomly chosen STS markers have already been used to screen the library, additional randomly chosen markers are increasingly likely to provide redundant information.

MAPPING WITH THE ENDS OF CLONED INSERTS

One alternative to random choice of STS markers is to generate the STS markers from one or both ends of the inserts in the mapping clones. There is a gain in efficiency of this strategy that comes from two sources. First, any STS marker is not positioned at random with respect to the cloned inserts but is known to be at the junction of at least one of them. For example, if an STS derived from the end of a cloned insert is used to screen a five-hit library, the resulting contig would be expected to include six clones (the clone from which the insert-end was determined plus five others) and extend an average of 1.8 times the length of the cloned insert.[15] (A contig of six clones is counterintuitive but results from the fact that the starting clone is known to be in the library and thus the distribution of clone numbers is Poisson with the zero term eliminated.) A second gain in efficiency from clone-based mapping is that selection of the clones for generation of the STS markers automatically biases the STS selection toward regions of the genome that are well covered with clones.

Even greater efficiencies are attained using both ends of the cloned inserts as STS markers. A pair of STS markers derived from the ends of a cloned insert define a region equal to the length of the insert already known to be in the library. The resulting contig extends the coverage of the region in both directions. A typical contig resulting from screening a 5-hit library with both STS markers would include 11 clones and would have an expected length of 2.6 times the length of the inserts in the clones.[15] Computer simulations discussed in the next section demonstrate quantitatively the gains in efficiency expected from such a double-end clone-limited strategy for STS content mapping.[15]

COMPARATIVE EFFICIENCY OF MAPPING STRATEGIES

The efficiencies of various STS selection strategies were compared by computer simulation.[15] Using the same simulation program, we have carried out simulations of three mapping strategies with parameters matching the *Drosophila* P1 system. The clone size is assumed to be 80 kb and the genome size 110 Mb. A library with a redundancy of coverage of 4.2 is assumed because this is the average number of clones recovered by screening the P1 library with random STS markers derived from sequences hit by the transposable element *P* (G. M. Rubin, personal communication). Coverage with a redundancy of 4.2 is expected to include 98.5% of the sequences in the genome. In all mapping strategies in the simulation, two clones are considered to overlap if and only if they hold one or more STS markers in common, and contigs are defined as groups of clones within which each clone overlaps with at least one other clone.

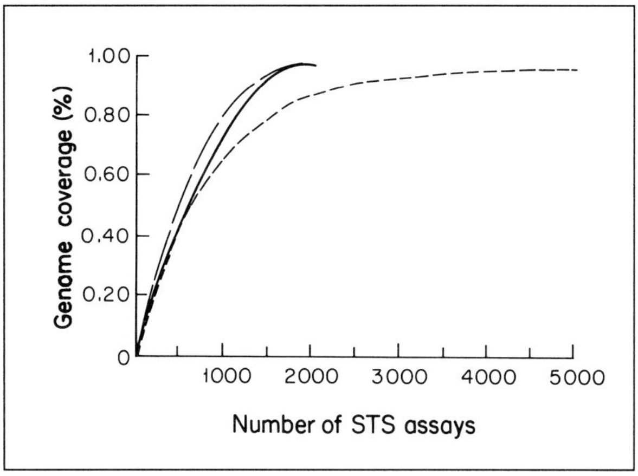

Fig. 6.5. Genome coverage as a function of number of STS assays for three STS-content mapping strategies. Solid curve, STS markers derived from both ends of cloned inserts; long dashes, STS markers derived from one end of cloned insert; short dashes, STS markers chosen at random positions in the genome. Data from a simulation program written by Palazzolo MJ, Sawyer S, Martin CH, Smoller DA, Hartl DL.Proc Natl Acad Sci USA 1991; 88:8034-8038.

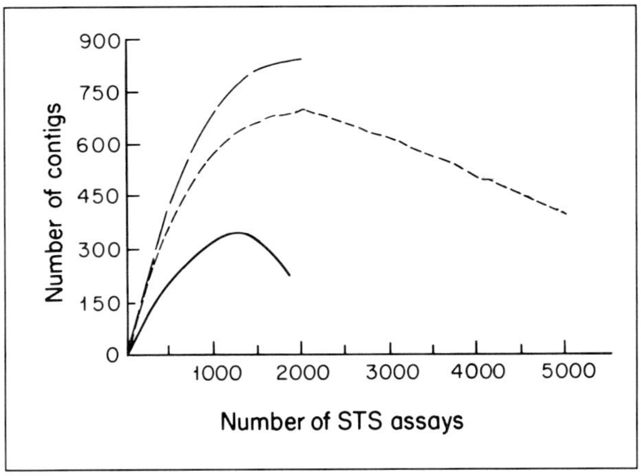

Fig. 6.6. Number of contigs as a function of number of STS assays for three mapping strategies. Solid curve, STS markers derived from both ends of cloned inserts; long dashes, STS markers derived from one end of cloned insert; short dashes, STS markers chosen at random positions in the genome. Data from a simulation program written by Palazzolo MJ, Sawyer S, Martin CH, Smoller DA, Hartl DL. Proc Natl Acad Sci USA 1991; 88:8034-8038.

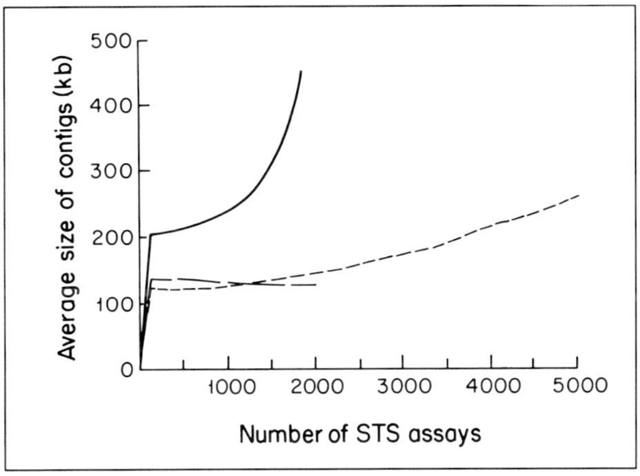

Fig. 6.7. Average size of contig as a function of number of STS assays for three mapping strategies. Solid curve, STS markers derived from both ends of cloned inserts; long dashes, STS markers derived from one end of cloned insert; short dashes, STS markers chosen at random positions in the genome. Data from a simulation program written by Palazzolo MJ, Sawyer S, Martin CH, Smoller DA, Hartl DL. Proc Natl Acad Sci USA 1991; 88:8034-8038.

Three approaches to STS content mapping were examined in the simulations. In the random-STS approach, each STS is chosen from a random position in the genome and used to screen the 4.2-hit library. In the single-end, clone-limited strategy, at each stage of the mapping, a new STS is generated from the end of an insert in a clone chosen at random from the library that has not previously been identified by screening with an earlier STS marker. Implementation of the double-end, clone-limited strategy is similar to the single-end, clone-limited method except that each clone chosen for STS generation has STS markers generated from both ends and these are used for screening the library. In both clone-limited approaches, the procedure is repeated until every clone in the library has either (1) been used to generate an STS or (2) has been hit by an STS. The maximum number of STS markers with clone-limited mapping is equal to the number of clones in the library (5,775 for the parameters chosen) but, in practice, most clones are assigned to contigs long before they would be chosen to generate STS markers.

The results of the simulations are summarized in Figures 6.5 through 6.7. The unbroken curves correspond to the double-end, clone-limited method. The curves in long dashes and short dashed are for the single-end, clone-limited strategy and for the random-STS markers, respectively. The curves for the clone-limited methods terminate at the point when all clones in the library have been either used for STS generation or have been assigned to contigs.

Figure 6.5 shows the progress of covering the genome with increasing numbers of STS assays. For small numbers of STS assays, there is little difference between the approaches. Initially, the single-end strategy is actually better for coverage than the double-end strategy because, at the beginning, each STS generates a different contig and, although the contigs are smaller, there are more of them. Eventually the double-end strategy catches up as the contigs begin to coalesce. By this point the inefficiency of random STS selection is quite clear: it requires 3,000 more STS assays to achieve the same overall coverage as with either of the clone-limited methods. The principal cause of the inefficiency of random STS selection is the tendency of STS markers chosen late in the experimental program to provide redundant information.

Comparison of the number of contigs resulting from the three mapping approaches are shown in Figure 6.6. Random STS selection yields an increasing number of contigs until a point at which about 2,000 STS assays have been carried out, after which the number of contigs starts to decrease because smaller, separate contigs found to hold an STS in common are brought together. Contig number is somewhat greater with single-end STS content mapping, but the procedure runs out of clones before there is a significant coalescence of the contigs. Making progress beyond this point requires a change in strategy. The superiority of the double-end strategy in contig generation is clearly apparent in Figure 6.6; there are never more than about 350 contigs during the entire mapping process.

Not only are there fewer contigs from the double-end STS strategy, the contigs are very much larger (Fig. 6.7). After about 1,500 STS assays have been carried out, the average size of the contigs with the double-end method is approximately 460 kb. The average contig size with random STS markers never gets above about 250 kb—even after 5,000 STS assays have been carried out. Perhaps counterintuitively, the single-end strategy performs the worst in generating large contigs. Indeed, after about 200 STS assays have been carried out, the average contig size decreases somewhat because new STS markers are drawn disproportionately from regions of the genome that are less well covered with clones.

Overall, the simulations imply that the double-end, clone-limited mapping strategy is some threefold to fourfold more efficient than random STS mapping in generating large contigs with nearly complete genome coverage. Specifically, for realistic parameters corresponding to P1 coverage of the *Drosophila* genome, it would require approximately 750 to 1,000 P1 clones (1,500 to 2,000 STS assays) to achieve greater than 98% coverage of the euchromatic genome in approximately two hundred contigs of average size at least 450 kb. In contrast, the method of random STS selection would require approximately 60% of the entire mapping effort to complete the final 20% of the physical map.

P1 CONTIGS BY STS CONTENT MAPPING

Implementation of the double-end, clone-limited mapping strategy reduces the time and cost required to build a high quality physical map. However, the random STS method offers some advantages in the early stages of mapping because the STS markers are generated independently of the clones. This consideration is particularly relevant to mapping the *Drosophila* genome because the random STS strategy can take advantage of the large number of *Drosophila* genes already cloned and sequenced. Since many of these genes have been particularly well mapped both genetically and physically, they provide a series of landmarks correlating the two types of genetic maps. Another large number of random STS markers can be generated from genomic sequences flanking recessive-lethal insertions of the transposable element *P*,[24-26] affording immediate access to many previously unanalyzed essential genes. The practical advantages of maximizing the biological content of physical maps have been emphasized repeatedly.[16,27,28]

The *Drosophila* genome also offers substantial improvements over random choice of clones used in the double-end mapping strategy. Because there is already a framework map based on in situ hybridization, physical mapping can be targeted to specific regions that are of particular interest in the earliest stages of genomic sequencing or for other reasons. In the following we describe the screening strategy and protocols in some detail.

POOLING STRATEGY

The working P1 library consists of 9,216 clones arrayed in 96 microtiter plates of 96 wells arranged in 8 rows and 12 columns. DNA for PCR screening was prepared at the Lawrence Berkeley

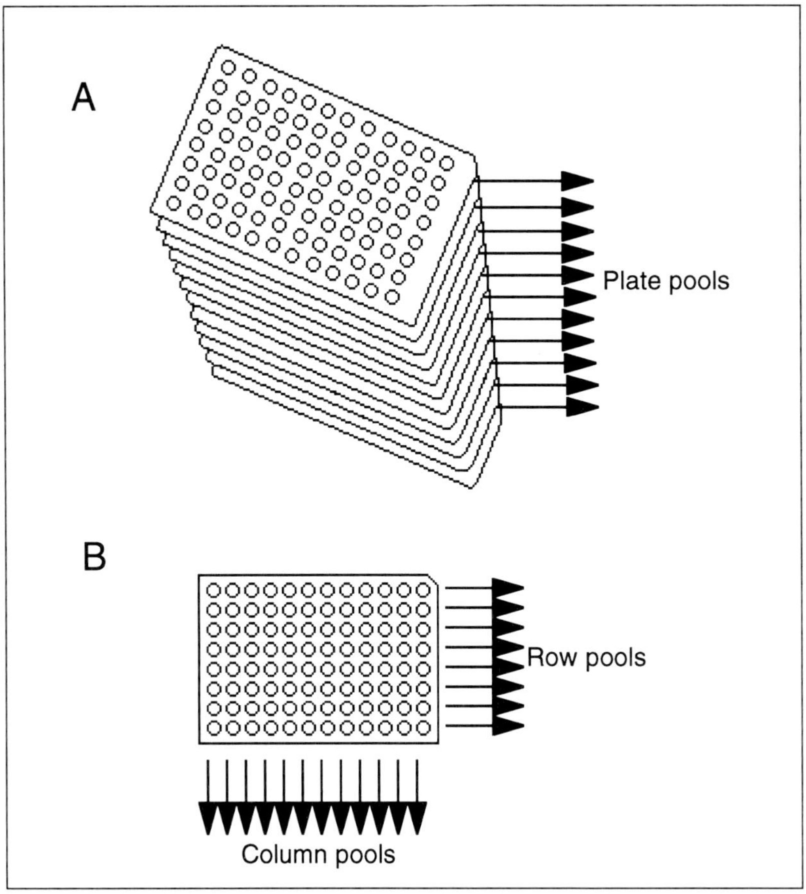

Fig. 6.8. DNA pooling strategy for PCR screening. A. Plate pools contain a mixture of DNA isolated from the 96 clones present in each plate. B. Row pools for each plate contain a mixture of DNA from the 12 clones in each row; column pools for each plate contain a mixture of DNA from the 8 clones in each column.

Laboratories according to the pooling strategy outlined in Figure 6.8. The PCR screening is carried out in two stages or "levels." In the first-level screen, 96 PCR reactions are carried out with "plate pools;" each plate pool contains the combined DNA from all 96 clones in one of the microtiter plates in the library. Once the plates containing the relevant clones have been identified in the first-level screen by their ability to support STS amplification, the second-level screen is carried out. For each plate subjected to a second-level screen, PCR assays are carried out on 8 "row pools," containing pooled DNA from each row, and 12 "column pools" containing pooled DNA from each column. Amplification will be supported by row pool *i* and column pool *j* only if the clone at coordinates *(i, j)* contains the STS. Ambiguity in clone identification can arise if a plate includes two (or more) clones containing the STS because, for example, clones at coordinates *(i, j)* and *(k, l)* will result in positive tests for row pools *i* and *k* and column pools *j* and *l* without identifying whether the positive clones are *(i, j)* and *(k, l)* or *(i, l)* and *(k, j)*. Resolution of the ambiguity requires PCR tests of the individual clones.

PCR SCREENING METHOD

For screening the DNA pools, the individual PCR reactions are carried out in 20 µl reaction volumes of PCR cocktail overlaid with mineral oil.[29] The concentration of $MgCl_2$ in the cocktail ranges from 1 to 3 mM, optimized for each pair of oligonucleotides. The PCR conditions are 25 cycles, each consisting of 3 seconds at 96° C, 45 seconds at 92° C, 90 seconds at the optimal annealing temperature for each pair of oligonucleotides, and 90 seconds at 72° C. The duration of time at the annealing temperature is increased by 1 second in each cycle, and the duration of time at 72° C is increased by 3 seconds in each cycle. The reactions are terminated by holding at 72° C for 5 minutes and stored at 4° C. PCR products are fractionated by gel electrophoresis in 1.2% agarose.

RECOVERY AND SEQUENCING THE ENDS OF P1 INSERTS

There are two major limitations to clone-limited mapping strategies. First, efficient mapping requires that both ends of the clones be used as STS markers, because mapping strategies based on single ends are only marginally better than random STS selection (Figs.s 6.6 and 6.7). Secondly, there must be an efficient method for determining the DNA sequences of large numbers of clone ends, otherwise much of the theoretical advantage of double-end, clone-limited mapping is offset by increased difficulty and expense of determining the clone-end DNA sequences.

The problem of recovering the ends of cloned inserts has been overcome in the P1 cloning system. Owing to the relatively high yield of cloned DNA, the vector-insert junctions can often be sequenced directly. An alternative method has also developed for specifically amplifying the vector-insert junction by PCR.[30] The amplification method makes use of a partially double-stranded oligonucleotide (the "adapter") with a single-stranded overhang complementary to a particular restriction site expected to be present in the insert; the adapter is used in combination with a second oligonucleotide specific to either the T7 or SP6 promoter flanking the cloning site of the P1 vector pAd10 sacBII diagrammed in Figure 2.2. The PCR adapter is ligated onto a restriction site in the cloned fragment near the junction of vector and insert, and a template suitable for PCR amplification is produced by primer extension from a site in the flanking vector sequence. In a test of the method using 92 P1 clones containing inserts of *Drosophila* DNA, both ends of the insert could be amplified in 76 clones (83% success rate). The procedures are rapid and can be carried out routinely and on a sufficiently large scale to be applicable to physical mapping of complex genomes.

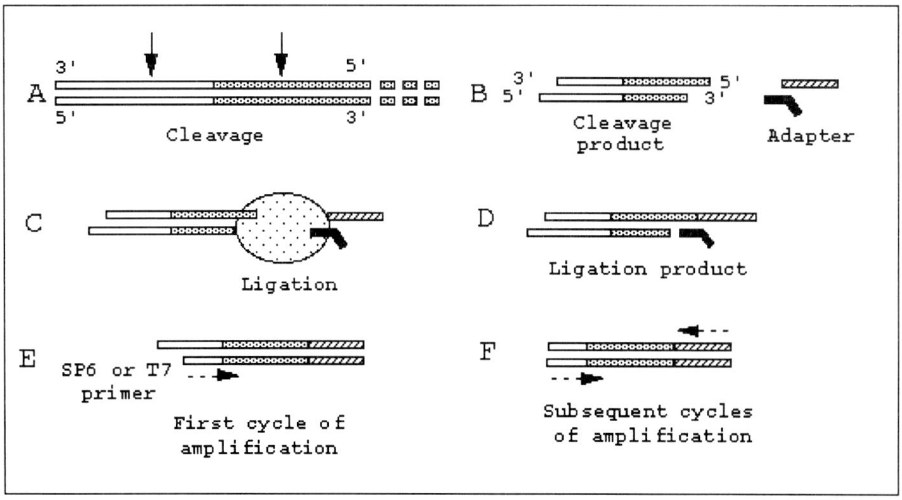

Fig. 6.9. PCR method for amplification of the ends of inserts in P1 clones. Nurminsky DI, Hartl DL. Biotechniques 1993; 15:201-208.

An outline of the method is shown in Figure 6.9. Experimental details are provided in the Appendix. In Figure 6.9, the open and stippled rectangles represent vector and insert DNA, respectively. In part A, P1 clones are cleaved with any of a number of restriction enzymes that create a 5'-CG overhang. The cleavage liberates the junction nearest either the SP6 or the T7 promoter (part B). In part C, a PCR adapter with a complementary 5'-CG overhang is ligated onto the ends. The shorter oligonucleotide of the PCR adapter (shown in black) is not phosphorylated at the 5' end to prevent its ligation to the cleavage product; the shorter oligonucleotide also contains three mismatched bases at the 3' end to prevent it from serving as a primer in the subsequent amplifications. For clarity, part D shows the adapter ligated only with the genomic end of the restriction fragment, but it also ligates with the vector end. After the ligation, the reaction mixture is divided into two parts to amplify the SP6 and T7 ends separately. To create a sequence suitable for PCR in the adapter, a vector primer specific for the relevant promoter is extended through the adapter (part E). In part F, the vector primer and an adapter primer are used for 25 cycles of PCR to amplify the genomic sequence of interest. After amplification of the genomic fragment, the DNA sequence is determined with an automated DNA sequencer and used to design oligonucleotides that are first tested on genomic DNA to ensure amplification of a single fragment of the expected size and then used in PCR assays of the entire P1 library to identify overlapping clones.

CONTIG GENERATION: AN ILLUSTRATION

Even as the P1 framework map based on in situ hybridization was being completed, contig assembly was initiated using the strategy of STS-content mapping. Because it is advantageous to correlate the physical map with biologically relevant DNA sequences, the mapping strategy has emphasized STS markers generated from known gene sequences available in GenBank and sites of insertion of the transposable element *P* as well as the double-end, clone-limited approach.[28] Here we wish to emphasize that most of the double-end, clone-limited P1 mapping is being carried out by the group at the Lawrence Berkeley Laboratory under the auspices of the *Drosophila* Genome Center, headed by G. M. Rubin. Substantial contigs have been generated through targeted mapping in the

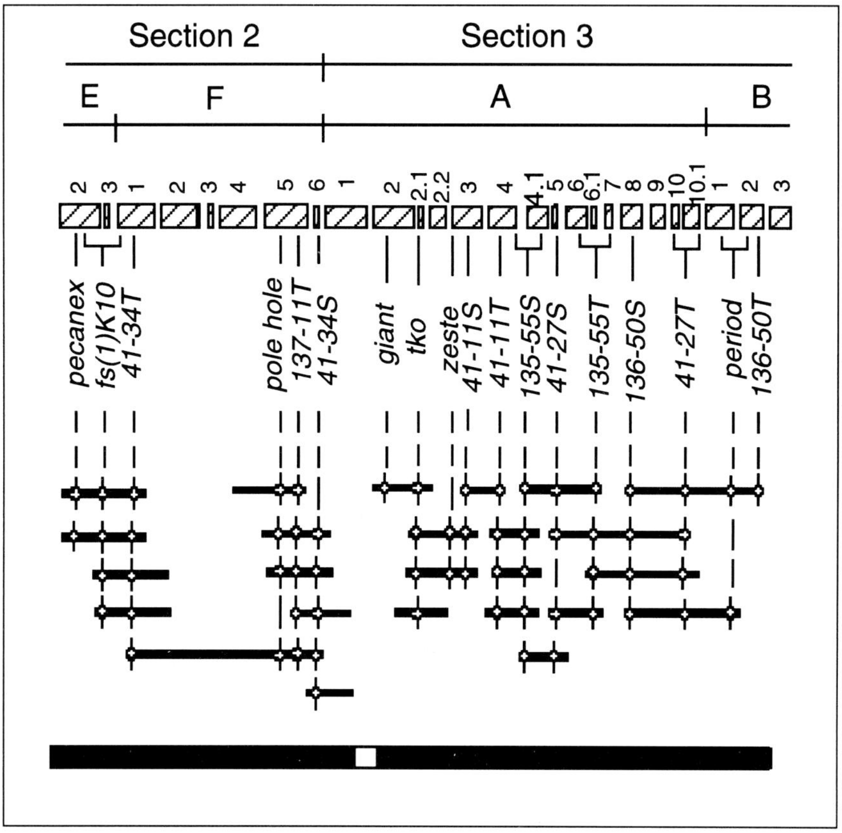

Fig. 6.10. Contigs in 2F;3A at the tip of the X chromosome. Hartl DL, Nurminsky DI, Jones RW, Lozovskaya ER. Proc Natl Acad Sci USA 1994; 91:6824-6829.

regions of *alcohol dehydrogenase* (section 35), the *Bithorax* complex (section 89), and the *Antennapedia* complex (section 84). In our laboratory in Cambridge we have also assembled large contigs in sections 1 through 3 of the X chromosome.

Although contig assembly is still in its early stages, the general strategy is illustrated by one of the targeted regions near the tip of the X chromosome shown in Figure 6.10. The hatched bars across the top represent the bands in the polytene chromosome scaled to the estimated amount of DNA.[13] The positions of the STS markers are indicated by vertical dashed lines intersecting with thin horizontal black bars below representing P1 clones; the ability of a P1 clone to support amplification of an STS is indicated with a cross at the site of intersection. The P1 contigs are indicated by thick horizontal black bars across the bottom. In this region, the

STS markers include 7 known genes and 11 termini of P1 inserts (the P1 inserts have hyphenated names) over approximately 650 kb. There is a contig of 10 clones extending approximately 240 kb from bands 2E2 to 3A1 and another contig of 15 clones extending approximately 370 kb from 3A1 to 3B3. Between these contigs there may be a small gap of no greater than 40 kb; screening to close this gap has not yet been carried out. The density of STS markers per P1 clone in this region is 2.5.

Recovery and Sequencing the Ends of YAC Inserts

In principle, a double-end, clone-limited strategy would also be effective for generating YAC contigs except that recovering the ends of the cloned inserts is somewhat more troublesome. With YACs, the only clone end that is readily recovered is the end nearest the centromere owing to its proximity to the bacterial origin of replication and ampicillin-resistance gene.[31,32] There are a number of methods for isolating the other end by various modifications of PCR..[32-35]

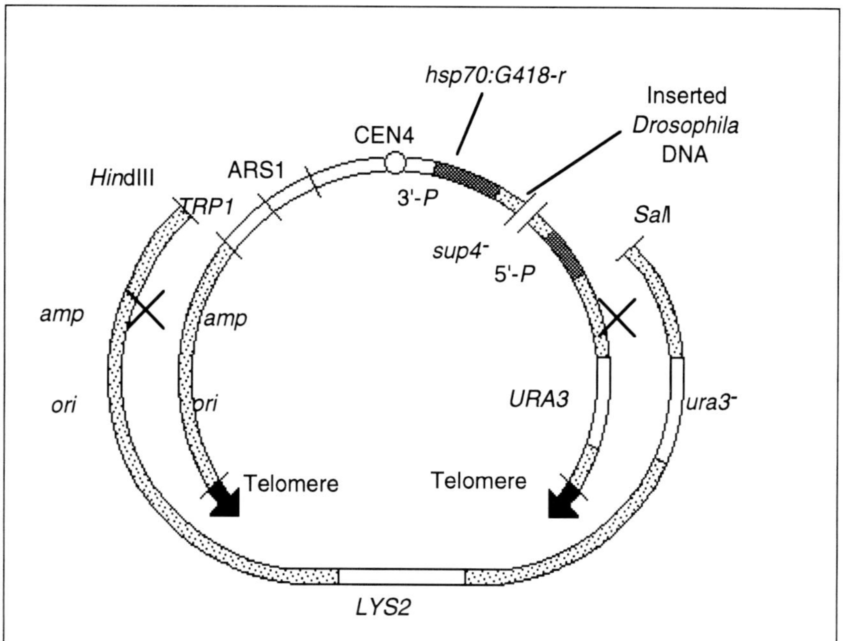

Fig. 6.11. Recombination events (crosses) needed to circularize a YAC clone with a linear DNA fragment bearing the LYS2 gene.

Circularization of YAC Clones

In some cases it is convenient to convert YAC clones into circular molecules, which can be achieved by homologous recombination in yeast by transformation with a linear *Hind*III-*Sal*I fragment bearing the *LYS2* gene from plasmid pPM680.[36] The principle of the method is illustrated in Figure 6.11. After transformation of a YAC-bearing yeast strain with the linear fragment, recombination between the fragment and homologous sequences near the vector ends results in the elimination of the terminal YAC vector sequences and the formation of a circularized clone (Fig. 6.12). Transformants are selected for lysine prototrophy and subsequently screened for uracil auxotrophy. YAC derivatives with the desired structure can be detected because the circularization alters the size

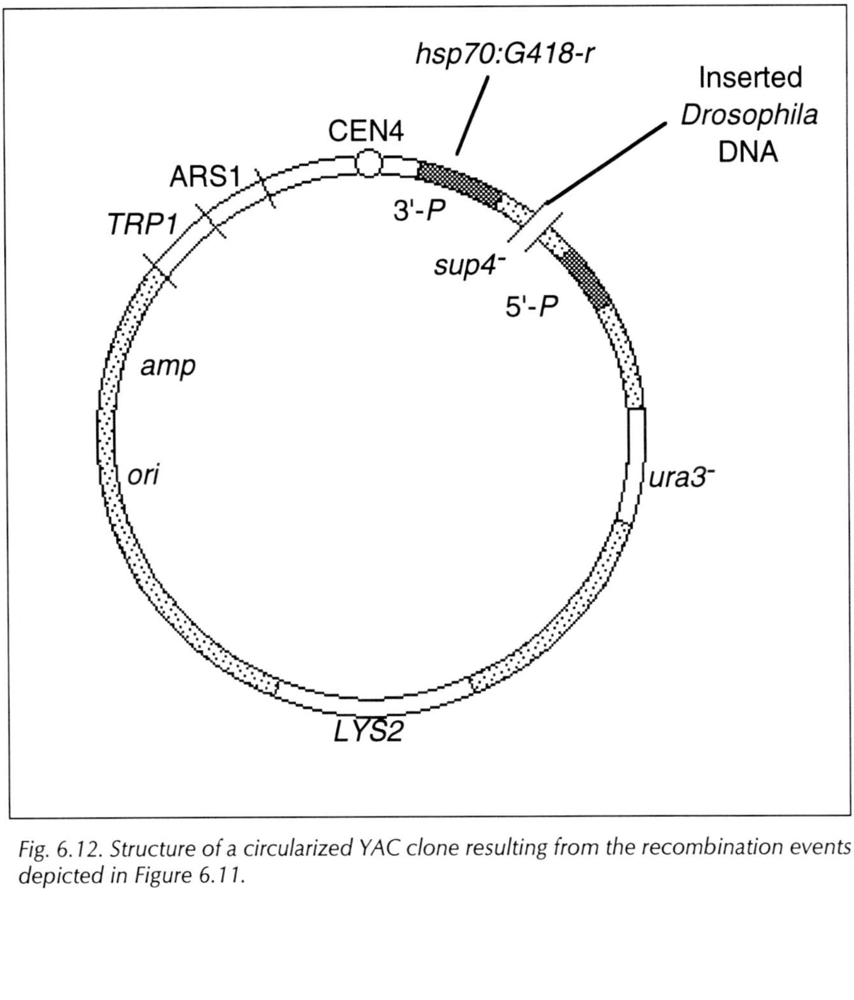

Fig. 6.12. Structure of a circularized YAC clone resulting from the recombination events depicted in Figure 6.11.

of certain diagnostic restriction fragments. For example, an *Eco*RI fragment of approximately 9 kb is diagnostic for all linear DY chromosomes derived from pYAC-P1, whereas a fragment of 12.5 kb is generated from the circularized derivatives of these clones. The relative sizes of *Bam*HI fragments also differ, with that from the left arm of the linear chromosome slightly smaller than the analogous *Bam*HI fragment from the circularized version; and that from the right arm of the linear chromosome approximately 3 kb smaller than the fragment generated from the circularized clone.[36]

REFERENCES

1. Olson MV, Dutchik JE, Graham MY, Brodeur GM, Helms C, Frank M, MacCollin M, Scheinman R, Frank T. Random-clone strategy for genomic restriction mapping in yeast. Proc Natl Acad Sci USA 1986; 83:7826-7830.
2. Kohara Y, Akiyama K, Isono K. The physical map of the whole *E. coli* chromosome: Application of a new strategy for rapid analysis and sorting of a large genomic library. Cell 1987; 50:495-508.
3. Coulson A, Sulston J, Brenner S, Karn J. Toward a physical map of the genome of the nematode *Caenorhabditis elegans*. Proc Natl Acad Sci USA 1986; 83:7821-7825.
4. Lehrach H, Drmanac R, Hoheisel J, Larin Z, Lennon G, Monaco AP, Nizetic D, Zehetner G, Poustka A. Hybridization fingerprinting in genome mapping and sequencing. In: K. E. Davies, S. M. Tilghman, eds. *Genome Analysis Volume 1: Genetic and Physical Mapping*. Cold Spring Harbor, NY: Cold Spring Harbor Laboratory Press 1990; 39-81.
5. Sidén-Kiamos I, Saunders RDC, Spanos L, Majerus T, Treanear J, Savakis C, Louis C, Glover DM, Ashburner M, Kafatos FC. Towards a physical map of the *Drosophila melanogaster* genome: Mapping of cosmid clones within defined genomic divisions. Nucleic Acids Res 1990; 18:6261-6270.
6. Riles L, Dutchik JE, Baktha A, McCauley BK, Thayer EC, Leckie MP, Braden VV, Depke JE, Olson MV. Physical maps of the six smallest chromosomes of *Saccharomyces cerevisiae* at a resolution of 2.6 kilobase pairs. Genetics 1993; 134:81-150.
7. Coulson A, Waterston R, Kiff J, Sulston J, Kohara Y. Genome linking with yeast artificial chromosomes. Nature 1988; 335: 184-186.
8. Hartl DL. Genome map of *Drosophila melanogaster* based on yeast artificial chromosomes. In K. E. Davies, S. M. Tilghman, eds. *Genome Analysis Volume 4: Strategies for Physical Mapping*. Cold Spring Harbor, NY: Cold Spring Harbor Laboratory Press 1992; 39-69.

9. Cai H, Kiefel P, Yee J, Duncan I. A yeast artificial chromosome clone map of the *Drosophila* genome. Genetics 1994; 136: 1385-1401.
10. Arratia R, Lander ES, Tavare S, Waterman MS. Genomic mapping by anchoring random clones: A mathematical analysis. Genomics 1991; 11:806-827.
11. Barillot E, Dausset J, Cohen D. Theoretical analysis of a physical mapping strategy using random single-copy landmarks. Proc Natl Acad Sci USA 1991; 88:3917-3921.
12. Ewens WJ, Bell CJ, Donnelly PJ, Dunn P, Matallana E, Ecker JR. Genome mapping with anchored clones: Theoretical aspects. Genomics 1991; 11:799-805.
13. Heino TI, Saura AO, Sorsa V. Maps of the salivary gland chromosomes of *Drosophila melanogaster. Drosophila* Information Service 1994; 73:619-738.
14. Beermann W. Chromomeres and genes. In: W Beermann, J Reinert, H. Ursprung, eds. *Developmental Studies on Giant Chromosomes.* New York: Springer-Verlag 1972; 1-33.
15. Palazzolo MJ, Sawyer S, Martin CH, Smoller DA, Hartl DL. Optimized strategies for sequence-tagged-site selection in genome mapping. Proc Natl Acad Sci USA 1991; 88:8034-8038.
16. Merriam J, Ashburner M, Hartl DL, Kafatos FC. Toward cloning and mapping the genome of *Drosophila*. Science 1991; 254:221-225.
17. Green ED, Olson MV. Chromosomal region of the cystic fibrosis gene in yeast artificial chromosomes: A model for human genome mapping. Science 1990; 250:94-98.
18. Brownstein BH, Nelson DL. *YAC Libraries: A User's Guide.* New York: W. H. Freeman 1994.
19. Ajioka JW, Smoller DA, Jones RW, Carulli JP, Vellek AEC, Garza D, Link AJ, Duncan IW, Hartl DL. *Drosophila* genome project: One-hit coverage in yeast artificial chromosomes. Chromosoma 1991; 100:495-509.
20. Craig AG, Nizetic D, Hoheisel JD, Zehetner G, Lehrach H . Ordering of cosmid clones covering Herpes simplex virus type I (HSV-I) genome: a test case for fingerprinting by hybridization. Nucleic Acids Res 1990; 18:2653-2660.
21. Evans GA. Combinatoric strategies for genome mapping. BioEssays 1991; 13:39-44.
22. Olson MV, Hood L, Cantor C, Botstein D. A common language for physical mapping of the human genome. Science 1989; 245:1434-1440.
23. Cohen D, Chumakov I, Weissenbach J. A first-generation physical map of the human genome. Nature 1993; 366:698-701.
24. Cooley L, Kelley R, Spradling A. Insertional mutagenesis of the *Drosophila* genome with single *P* elements. Science 1988; 239: 1121-1128.

25. Cooley L, Berg C, Spradling A. Controlling *P* element insertional mutagenesis. Trends Genet 1988; 4:254-258.
26. Berg CA, Spradling AC. Studies on the rate and site-specificity of *P* element transpositions. Genetics 1991; 127:515-524.
27. Kafatos FC, Louis C, Savakis C, Glover DM, Ashburner M, Link AJ, Sidén-Kiamos I, Saunders RDC. Integrated maps of the *Drosophila* genome: progress and prospects. Trends Genet 1991; 7:155-161.
28. Hartl DL, Palazzolo MJ. *Drosophila* as a model organism in genome analysis. In: K. W. Adolph, eds. *Genome Research in Molecular Medicine and Virology.* Orlando, FL: Academic Press 1993; 115-129.
29. Hartl DL, Nurminsky DI, Jones RW, Lozovskaya ER. Genome structure and evolution in *Drosophila:* Applications of the framework P1 map. Proc Natl Acad Sci USA 1994; 91:6824-6829.
30. Nurminsky DI, Hartl DL. Amplification of the ends of DNA fragments cloned in bacteriophage P1. Biotechniques 1993; 15:201-208.
31. Burke DT, Carle GF, Olson MV. Cloning of large segments of exogenous DNA into yeast by means of artificial chromosome vectors. Science 1987; 236:806-812.
32. Garza D, Ajioka JW, Carulli JP, Jones RW, Johnson DH, Hartl DL. Physical mapping of complex genomes. Nature 1989; 340:577-578.
33. Ochman H, Gerber AS, Hartl DL. Genetic applications of an inverse polymerase chain reaction. Genetics 1988; 120:621-623.
34. Ochman H, Ajioka JW, Garza D, Hartl DL. Inverse polymerase chain reaction. In: H. A. Erlich, eds. *PCR Technology: Principles and Applications for DNA Amplification.* New York: Stockton Press 1989; 105-111.
35. Riley J, Butler R, Ogilvie D, Finniear R, Jenner D, Powell S, Anand R, Smith JC, Markham AF. A novel, rapid method for the isolation of terminal sequences from yeast artificial chromosome (YAC) clones. Nucleic Acids Res. 1990; 18:2887-2890.
36. Garza D, Ajioka JW, Burke DT, Hartl DL. Mapping the *Drosophila* genome with yeast artificial chromosomes. Science 1989; 246:641-646.

CHAPTER 7

GENOMIC SEQUENCING

A clone-based physical map consists of a set of ordered, over-lapping inserts of cloned genomic DNA. Who needs it? True enough, a physical map is not necessary either to induce mutations or to map them genetically. In *Drosophila*, because of the polytene chromosomes, a physical map is not even necessary to clone genes for those hardy souls not afraid of a little chromosome walk. However, even in *Drosophila*, many chromosome walks get hopelessly mired in repetitive DNA, and so one use of a clone-based physical map is to make cloning genes easier and more reliable. More importantly, a clone-based physical map provides ready access to any region of the genome in the form of clones that can be used as substrates for genomic sequencing. Furthermore, it is quite clear, even from the first eukaryotic chromosome completely sequenced, that many open reading frames can be rendered nonfunctional without producing any detectable phenotype.[1] Either many mutations have such subtle phenotypes that they escape detection or the genes have overlapping functions with other genes and so are redundant. In either case a significant fraction of genes will be missed by a strategy of genetic analysis driven solely by the acquisition of new mutations having detectable phenotypic effects.

To go beyond the limitations of conventional mutagenesis a genome-wide strategy is needed. One such genome-wide strategy is the enhancer-trap method of identifying insertions of a *P*-element transposon according to the pattern of expression of a reporter gene induced by stage- or tissue-specific enhancers near the insertion site.[2-4] The method is independent of the phenotype of mutations in the target gene because the phenotypic assay is for the effect of enhancers of the target gene on the expression of the reporter gene in the transposon. However, a limitation of enhancer trapping is the nonrandom distribution of *P*-element insertion sites, which can be overcome in part by using a different transposable element as the basis of the system, such as the *hobo* element.[5]

A second genome-wide strategy, independent of the phenotype of mutations, is genomic sequencing. This approach is comprehensive and in principle unbiased but requires a clone-based physical map for its implementation. The P1 map of the *Drosophila* euchromatin has become the primary resource for genomic sequencing. However, even before the P1 map became available, *Drosophila* was one of the principal model organisms singled out for complete genomic sequencing under the auspices of the Human Genome Project, along with *E. coli*, *S. cerevisiae*, *C. elegans*, and *M. musculus*.[6,7] Because the P1 map of *Drosophila* covers only the euchromatin, there is a bias in representation. Eventually, one hopes, the genetically interesting parts of the heterochromatin will also be mapped and sequenced, and now that an enhancer-trap method of identifying expressed genes in heterochromatin is available,[4] this prospect seems quite likely to be feasible.

WHERE ARE WE NOW?

Since GenBank was created in the early 1980s, the number of base pairs in the database has double approximately every 21 months. As of GenBank Release 81.0 (Spring, 1994) the top ten species with the most sequence deposited are shown in Table 7.1.

Table 7.1. Extensively sequenced genomes

Species name	Organism	Total reported sequence (megabases)
Homo sapiens	Human	33
Mus musculus	Mouse	12
Rattus norvegicus	Rat	7.3
Saccharomyces cerevisiae	Yeast	6.4
Escherichia coli	Bacteria	5.2
Drosophila melanogaster	Fruit fly	4.9
Caenorhabditis elegans	Nematode	4.9
Arabidopsis thaliana	Wall cress	2.8
Gallus gallus	Chicken	2.5
Oryza sativa	Rice	2.0

Very little reported sequence is from long, contiguous stretches of the genome. The sequences are mostly from individual genes, cDNAs, or multiple alleles in natural populations. (The latter explains why the *E. coli* sequences in GenBank exceed the genome size.) Most of the sequences, it should also be noted, were obtained by conventional, manual autoradiographic sequencing methods at a cost between $3 and $7 per base pair depending on the efficiency of the laboratory doing the work.

In the future, if the Human Genome Project has its way, an increasing amount of GenBank sequence will be in the form of long, continuous stretches of the genome. Although the goal of the Human Genome Project is to complete the sequence of the human genome and that of selected model organisms within the next 15 years, the timetable for sequencing the model organisms is much shorter. According to NIH Program Announcement PA-94-045, "the sequencing of the DNAs of those organisms is underway and the extended goals call for the completion of the *E. coli* and *S. cerevisiae* sequences by or before 1998, for the *C. elegans* sequence to be approaching completion by 1998, and for the *D. melanogaster* sequence to be approaching completion by 2000" (NIH Guide for Grants and Contracts, March 11, 1994).

Genomic sequencing is a significant challenge in terms of output and cost. The current maximum rate of sequencing of large contiguous stretches of DNA by any group is been approximately 2 Mb per year, but this can easily be scaled to 5 Mb or more by increasing the number of personnel and automated DNA sequencers within each major group. Another problem is cost: including supplies, labor, amortization of instruments, and overhead, the cost of finished DNA sequence is still approximately $0.75 per base pair. This amount is a significant improvement over the cost of sequencing by conventional, autoradiographic means and brings the total cost of sequencing model organisms (other than the mouse) into a reasonable range. Significant additional cost savings will, however, be required to make it financially feasible to achieve the goal of complete sequencing of the 3 billion bases in the human genome.

SHOTGUN SEQUENCING

The currently favored method for genomic sequencing is "shotgun" sequencing, in which clones of a suitable size are selected at random for sequencing.[8-10] The favored templates for shotgun sequencing are M13 (single stranded)[10] and pUC-based plasmids (double stranded).[11] An advantage of double-stranded templates is the possibility of obtaining sequence from both ends of the subcloned inserts.

Assuming an equal probability of subcloning across any large region of interest, generation of a sufficient amount of random sequence must eventually cover the entire target. Complete shotgun coverage requires exquisitely random subcloning and highly redundant sequencing, however, and so in practice there is a stopping point short of complete coverage at which the existing sequences are assembled into contigs and the process of "closure" begun. In the closure process, additional sequencing is carried out to close gaps and to cover the second strand in regions across which only single-stranded sequence was obtained from shotgunning. Closure methods are ad hoc, sometimes based on resequencing templates that were unsatisfactory in the shotgun phase and often based on the synthesis of oligonucleotides to allow "primer walking" further along templates known to contain gaps in the sequence to generate PCR products to close gaps. Successful application of shotgun sequencing includes large stretches of the *E. coli*[12] and *C. elegans*[13,14] genomes.

Advantages of Shotgun Sequencing

Shotgun sequencing offers several advantages. First, nothing need be known about the target sequence in order to subclone and begin sequencing. Secondly, most of the sequence generated in the early stages is nonredundant and so contributes maximally to the complete assembled sequence. Third, shotgun sequencing lends itself to an assembly-line organization and many steps can be automated or semiautomated. The use of double-stranded templates in shotgun sequencing has the additional advantage of simplifying closure in that the templates containing unsequenced regions can often be identified among those whose ends have already been sequenced; these templates can then be subjected to resequencing to obtain longer runs or primer walking to close the gaps.

It is interesting to observe how the technology of shotgun sequencing has developed. There have been no major breakthroughs in hardware or software resulting in sudden improvements of a factor of five or ten. Rather, the progress has come through incremental improvements of a factor of 50% here, 20% there, 30% somewhere else. When implemented together, however, the incremental improvements have aggregated to a substantial gain in throughput and decrease in cost over just a few years.

LIMITATIONS OF SHOTGUN SEQUENCING

There is, however, a sort of "Poisson catastrophe" in shotgun sequencing in which any new sequence generated is increasingly likely to be redundant with sequences obtained previously. For example, sequencing a 100 kb P1 clone with sixfold redundancy requires 600 kb of raw sequence of which approximately 70,000 nucleotides are represented between 4 and 8 times, approximately 2,000 nucleotides are represented 12 or more times, and about 250 nucleotides are not represented at all. Any new subclone is highly unlikely to cover any of the gaps. With a decrease in average redundancy to four, the number of nucleotides sequenced between 4 and 8 times decreases to approximately 55,000 and only about 90 nucleotides are sequenced 12 or more times; however, the decrease in redundancy comes at the cost of more gaps—in this example about 1,800 nucleotides not sequenced at all.

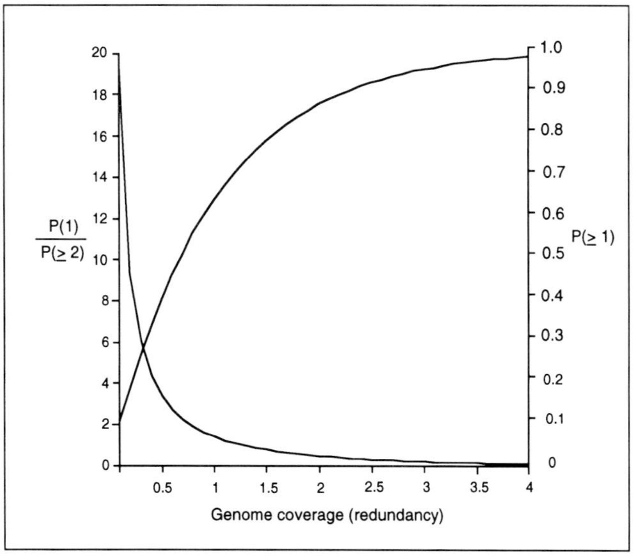

Fig. 7.1. The Poisson catastrophe. With increasing redundancy and Poisson coverage, the ratio of nucleotides hit once to those hit two or more times plummets whereas coverage of the target genome increases slowly. The left axis is for the descending curve, which follows $y = x\,e^{-x}/(1 - e^{-x} - x\,e^{-x})$. The right axis is for the ascending curve, which follows $y = 1 - e^{-x}$.

The extent of the Poisson catastrophe is plotted in Figure 7.1, in which the left axis and descending curve shows the ratio of the number of nucleotides sequenced once relative to the number sequenced two or more times as a function of increasing redundancy of coverage. For low levels of coverage, the ratio is very high (approaching 20) indicating that almost all of the sequence information is unique. Unfortunately, low redundancy also implies very incomplete coverage of the target genome. With one-hit redundancy, the ratio $P(1)/P(\geq 2)$ equals 1.39 and the genome coverage $P(\geq 1)$ equals 63%; with two-hit redundancy the corresponding numbers are $P(1)/P(\geq 2) = 0.46$ and 86% coverage; and with three-hit redundancy the corresponding numbers are $P(1)/P(\geq 2) = 0.19$ and 95% coverage. In other words, with two-hit redundancy about 68% of the sequence information is superfluous, and with three-hit redundancy about 84% is superfluous. Furthermore, the Poisson catastrophe depicts the best of all possible worlds in assuming consummate randomness in the subcloning. Any biases in the subcloning or sequencing only serve to make matters worse.

DIRECTED SEQUENCING

Much of the difficulty in genomic sequencing results from the fact that the long-range order of sequences present in the genome itself or in large-insert clones such as YACs, P1s, and cosmids is lost in the subcloning necessary to obtain templates for DNA sequencing. The long-range sequence organization is recovered only in the assembly and closure phases of shotgun sequencing. Contemplation of the Poisson catastrophe, as well as considerations of sequence assembly, make it clear that genomic sequencing would be more rapid, efficient, and require less redundancy if long-range sequence organization could be preserved throughout the analysis or recovered relatively easily from the sequencing templates.

Preservation of long-range sequence organization entails direct genomic sequencing or the direct sequencing of clones with large inserts. To date, no method has been developed for genomic sequencing that has the necessary accuracy, throughput, and cost required for application to complex genomes. Similar limitations apply to methods that might in principle be applicable to the direct sequencing of YACs, P1 clones, or cosmids, such as transposon-facilitated DNA sequencing.[15]

The alternative to preserving long-range sequence organization is to be able to recover it relatively easily from the sequencing templates. Such is the case when there is a detailed restriction map of the target sequence or a high-resolution STS map. However, the generation of such maps is itself a challenge unless it is incorporated into the sequencing method.

SPECTRUM PCR

In a method called "spectrum PCR," long-range sequence organization in the form of a high-resolution physical map is generated in the course of selecting templates for sequencing.[16] This method is particularly suited for sequencing clones with large inserts, such as cosmids or P1 clones. Initially, DNA from the target clone is sheared in a French press and fractionated by electrophoresis to isolate fragments approximating 3 kb. These fragments are purified, their ends repaired with Klenow and treated with phosphatase, and then ligated into a plasmid vector such as pSP72 (Promega). After transformation, 960 clones are selected at random and arrayed into 10 96-well microtiter dishes. DNA pools are prepared according to the row x column x plate format as outlined in Chapter 6 (see Figure 6.8).

Once the subclones and DNA pools are prepared, a set of sequencing templates is selected by a walking strategy using PCR screening, simultaneously with which the templates are also organized into a high-resolution physical map of the target. For each PCR screening, the primers consist of one oligonucleotide specific for a small sequence in the target and another oligonucleotide specific for the plasmid vector (such as pSP72) adjacent to the cloning site.

The walking strategy is outlined in Figure 7.2. The target sequence (for example, a P1 clone) is indicated by the bar across the top, with the direction of the desired walk going from left to right. The rectangles below represent individual 3-kb plasmid subclones flanked by the vector sequences A and B. For any known sequence in the target, indicated by the vertical line, two sets of PCR reactions are carried out with each pool of DNA. One set of reactions uses a primer specific to the known target sequence (left-pointing arrow) combined with a primer specific to the A side of the cloning site (right-pointing arrow); the other set of reactions uses the same target-sequence primer combined with a primer specific to the B side of the cloning site. For example, the A and B primers used with pSP72 are

5'-ATTTAGGTGACACTATAGAACTCGCGC-3'

(for the SP6 side) and

5'-GATGAATTCGAGCTCGGTACC-3'

(for the T7 side).[16]

Any subclone containing the target sequence will support amplification with either the A or B primer, depending on the orientation of the insert. Figure 7.2 shows 10 subclones containing the target sequence, five supporting amplification with the A primer and five with the B primer. Furthermore, the size of the PCR product indicates how far the insert in each subclone extends in the direction of the walk. With subclones having 3 kb inserts, for

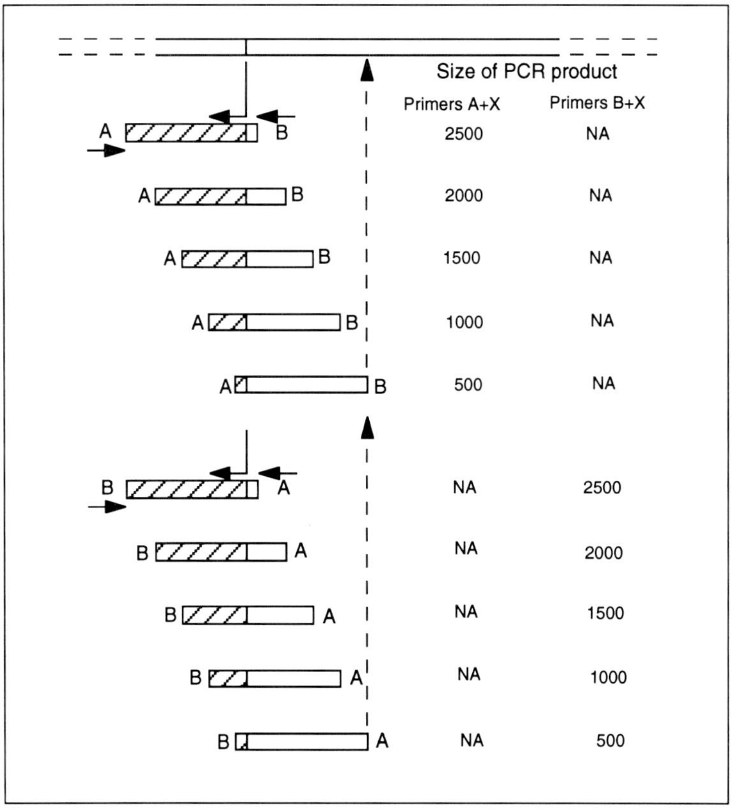

Fig. 7.2. Walking strategy in spectrum PCR. The bar across the top represents the target sequence and the rectangles below are 3 kb plasmid subclones flanked by vector sequences A and B. Hatched areas represent PCR products amplified using the primers indicated by the arrows. NA indicates no amplification.

example, a PCR product of 2.5 kb using the A primer indicates the target sequence is present in a subclone at the position and in the orientation shown in subclone 1, which includes 500 bp extending the walk in the direction of B. Such a subclone does not extend the walk very far. However, a subclone yielding a PCR product of, say, 500 bp must contain the target sequence in the position shown in subclones 5 and 10 in Figure 7.2, with the insert oriented either as in subclone 5 (amplified with primer A) or as in subclone 10 (amplified with primer B). In either case, the insert extends the walk 2.5 kb in the desired direction. Once a subclone of this type has been identified, the end of the insert in the direction extending the walk is sequenced (dashed vertical line),

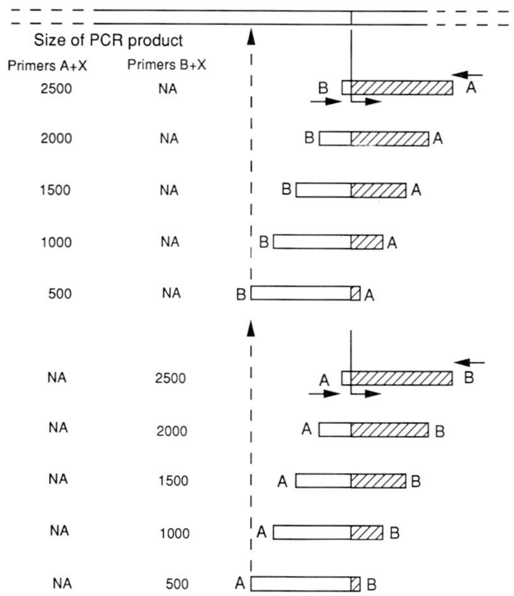

Fig. 7.3. Walking by spectrum PCR in the right-to-left direction. NA indicates no amplification.

and this sequence is used to design a synthetic oligonucleotide used as the target sequence to repeat the PCR assays and take another step. With what sequence do you start the walk? A logical choice is the sequence near the vector-insert junction in the original cosmid or P1 clone.

With spectrum PCR using 3 kb subclones and selection for 500 bp overlaps, approximately 32 steps are required to cover an 80-kb insert in a P1 clone. The procedure therefore requires 32 walking cycles, each of which includes carrying out the PCR reactions, sequencing the appropriate vector-insert junction, and synthesizing a new oligonucleotide. Hence, spectrum PCR requires an up-front commitment in terms of time and oligonucleotide synthesis, but the procedure yields a set of ordered, overlapping templates for sequencing as well as an STS approximately every 2.5 kb.

The walk across a large-insert clone by spectrum PCR can be speeded up by walking simultaneously in both directions. Figure 7.2 shows walking in the direction left to right, Figure 7.3 shows walking in the direction right to left. The procedures are identical except that the walking primers are synthesized to amplify opposite DNA strands. In this hypothetical example, ten subclones support amplification with the walking primer—five with the A end of the vector and five with the B end of the vector. The subclones used for the next step would be either the fifth or the tenth listed because the inserts in these subclones extend maximally in the direction of the walk. From either of these clones, the sequence at the insert end provides the information needed for synthesizing the next walking primer.

There is an important bookkeeping issue in regard to bidirectional walking with spectrum PCR: one must keep track of all of the subclones that support amplification with each walking primer as the walk progresses in both directions. The bookkeeping is necessary because the walk terminates when one or more subclones are identified that support amplification of walking primers going in opposite directions. Successful termination of the overall walk is achieved when the individual walks from each end meet somewhere in the middle in a common subclone.

Successful termination of a bidirectional walk in illustrated in Figure 7.4. In this hypothetical example the leftward and rightward walks come together in the identification of the two clones marked with asterisks, which support amplification of walking primers from both ends. This example represents an ideal case in which the inserts joining the walks from either end extend nearly maximally in the direction of the respective walks. An ideal case like this can take place only when the walking primers are separated by a distance slightly smaller then the size of the inserts in the subclones. Such fortuitous spacing is expected to be unusual, and so the clone or clones joining the walks will typically not be among those that would be chosen to continue the walks—hence the importance of good bookkeeping.

Given additional sequence data from the interior of the target molecule, walks can be initiated at multiple points and proceed simultaneously.[16] Bookkeeping becomes even more important in this approach so that the coalescence of the walks can be recognized efficiently. Although multiple simultaneous walks decrease the total time necessary for full coverage by spectrum PCR, the disadvantages are (1) the need for STS markers at multiple interior points in the target sequence, (2) the generation of some redundant data, and (3) a somewhat greater demand for oligonucleotide primers.

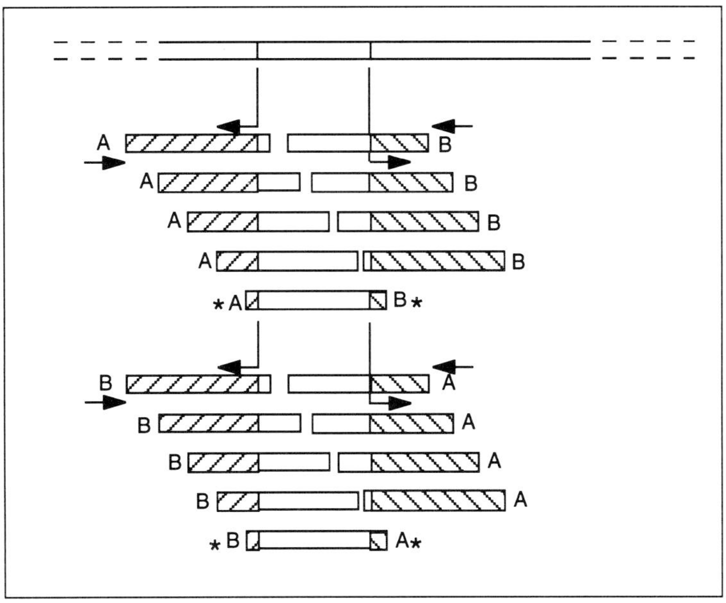

Fig. 7.4. A bidirectional walk with spectrum PCR terminates when a subclone somewhere in the middle is identified that contains both walking primers.

TRANSPOSON-FACILITATED SEQUENCING

Choice of 3 kb inserts for spectrum PCR was originally motivated by considerations of DNA sequencing using transposon insertions as entry points in the target sequence.[15] Transposon insertions into a 3-kb sequence can be mapped with a pair of PCR reactions, each amplifying the DNA sequence between the transposon and one of the vector-insert junctions. The two reactions serve as controls of one another because the sum of the amplified fragment lengths must equal the total insert size (for example, 3 kb). Although a size of 3 kb is approximately the length that can easily be amplified with conventional PCR methods, recent innovations allow reliable amplification of larger fragments.[17]

The purpose of mapping a sufficiently large number of transposon insertions by PCR is to identify a set of insertions with their sites of insertion suitably spaced to enable sequencing of virtually all of the target sequence from primers that anneal to the

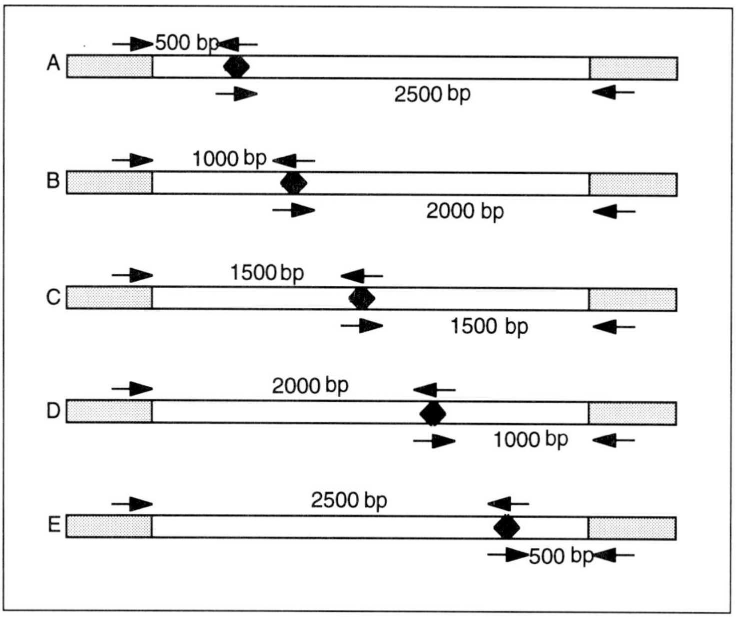

Fig. 7.5. Transposon-based sequencing of 3-kb templates isolated by spectrum PCR. In the first step, a large number of independent transposon insertions (black diamonds) are localized by estimating the sizes of the PCR products bridging between the transposon and vector sequences at either end of the insert. A set of subclones with suitably spaced inserts are selected as sequencing substrates, and sequencing is carried out with primers specific to the transposon ends. In this example, a set of sequencing templates with insertions every 500 bp has been selected.

transposon ends. A hypothetical example is the set of insertions illustrated in Figure 7.5, in which the insertion sites are evenly spaced at 500-bp intervals. A set of uniformly spaced insertions like those in Figure 7.5 would be selected from a much larger set after mapping the insertion sites. It is unfortunately the case that transposon insertions are typically nonrandom—sometimes grossly nonrandom—and a small number of "hotspots" can account for a disproportionate share of all insertion events. Nevertheless, there is no need for absolute randomness of insertion in order to select a set suitable for sequencing. There is, however, a requirement for a sort of quasirandomness in which there are many possible insertion sites spread widely throughout the target sequence with frequencies of insertion that are high enough for an acceptable set of sequencing clones to be selected by mapping a reasonable number of independent insertions.

With a set of insertions like that in Figure 7.5, sequencing reactions carried out in the directions given by the arrows, using primers specific either to the transposon ends or to the vector-insert junctions, are expected to cover overlapping regions and include the entire target sequence. In practice complete coverage is not achieved (in part because of the inevitable nonrandomness of transposon insertion) and a closure process is required in which gaps are filled with longer sequencing runs from particular templates or by oligonucleotide walking.

ADVANTAGES OF DIRECTED SEQUENCING

Directed sequencing has a number of advantages over pure shotgun sequencing. In a directed strategy, the templates to be sequenced are chosen in an orderly fashion so as to minimize sequence redundancy while still affording closure. The nonrandom choice of sequencing templates presupposes some knowledge of long-range organization of the target sequence in the form of a restriction map or an STS map of the sort generated by spectrum PCR. Such long-range sequence information is also useful in closure because it facilitates the relative localization of contigs and, if extensive enough, can be used to find the locations and approximate lengths of the gaps as well as to identify the templates that need to be sequenced to close the gaps.

Limitations of Directed Sequencing

The main limitation of directed sequencing is the initial investment of time and effort needed to select the templates used for sequencing. Restriction mapping a large insert is a very difficult and tedious job unsuitable for automation. Although there is also a certain tediousness in spectrum PCR in walking across an 80 kb insert in steps of 3 kb, many of the procedures are suitable for automation. The main limitations of spectrum PCR are the numbers of oligonucleotide primers that need to be synthesized and the occasional false positives in the PCR reactions that require sorting out.

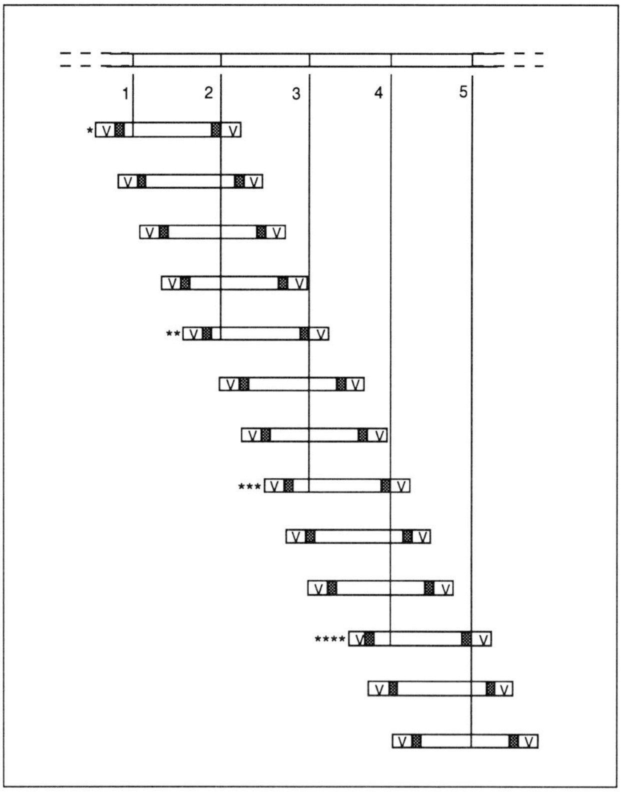

Fig. 7.6. Ordered shotgun sequencing makes use of a ordered set of 10-kb templates whose ends have been sequenced. The vertical lines numbered 1 through 5 indicate STS markers used to map the templates by long-range spectrum PCR. The insert in each subclone is illustrated as a rectangle, and the shading at the ends indicates regions of known sequence. The letter V denotes the vector ends flanking the cloning sites, which are not distinguished typographically although they do differ in sequence. The subclones chosen for walking at each stage in the process are indicated with asterisks.

Directed sequencing also includes an important data handling component because parallel processing of several independent clones requires careful bookkeeping. One can hardly contemplate transposon-facilitated sequencing on the scale required for large-scale genomic sequencing without automated imaging, data acquisition, and analysis in view of the large number of transposon insertions that need to be mapped.

ORDERED SHOTGUN SEQUENCING

As the Poisson catastrophe sets a practical limit to the utility of shotgun sequencing, excessive emphasis on procuring an ideal set of ordered templates erodes the efficiency of directed sequencing. One of the implications of Figure 7.1 is that shotgun sequencing is highly cost effective in the early stages when almost all of the sequence information is new. Its effectiveness drops precipitously, however, and so shotgun sequencing should not go on too long. In directed sequencing using spectrum PCR for the generation of an ordered set of sequencing templates and transposon insertion for the generation of an ordered set of sequencing substrates, there is a considerable effort expended in PCR mapping. While everything is precisely ordered by these procedures, it is a pity that the advantages of the early stages of shotgun sequencing are forfeited.

With gel-based sequencing, an optimal procedure probably includes components of both the shotgun and directed strategies. One approach for combining strategies that seems promising is outlined in Figure 7.6. For concreteness we will call this strategy "ordered shotgun sequencing." The diagram for ordered shotgun sequencing resembles that in Figure 7.2 for spectrum PCR, but there are some critical differences:

- The subclones in Figure 7.2 have considerably longer inserts than in spectrum PCR—for example, 10 kb. The longer insert size means that conventional PCR will not usually amplify from either vector-insert junction to the position of an STS somewhere in the middle, and long-range PCR must be used instead.[17]
- Because the subclones are longer, fewer of them are needed to cover a target clone with suitable redundancy. For example, 96 subclones with inserts of 10 kb are needed to cover an 80-kb P1 insert with 12-fold redundancy. (With

large inserts, it is worth the effort to screen out subclones containing excessive lengths of vector sequence, either by hybridization or with PCR.) The smaller number of subclones eliminates the need for pooling, and so the PCR amplifications can be carried out with single templates.

- In ordered shotgun sequencing, all of the vector-insert junctions of the subclones are sequenced; this is the "shotgun" component of the procedure, and it is maximally efficient because virtually all of the sequence is useful. For example, 350 bp of sequence from each end of 96 subclones totals 67.2 kb, which is equivalent to coverage of an 80-kb target sequence with a redundancy of 0.84; increasing the length of the sequencing runs to, say, 550 bp increases the redundancy to 1.32. Shotgun sequencing is still very efficient at these levels of redundancy (Fig. 7.1).
- In spectrum PCR subclones not used for walking are discarded; in ordered shotgun sequencing, the STS at each end of each clone is ordered and spaced relative to neighboring ends in the process of walking and this information is incorporated into the finished sequence.

The procedure of ordered shotgun sequencing therefore requires a set of 96 subclones with approximately 10 kb inserts. Ideally, the set would not include subclones consisting primarily of vector sequence, and the relative uniformity of the insert sizes would have been verified by long-range PCR. Each of the 96 subclones is used as a template for sequencing from each end, and the end sequences are assembled into such contigs as may exist at this stage.

Then the subclones are mapped as illustrated in Figure 7.6. The first STS (vertical line number 1) is employed as a PCR primer in combination with each of two different vector-specific primers flanking the cloning site. In Figure 7.6, the vector ends are both denoted "V" for simplicity, but they are distinguishable. Subclones containing an insert with the STS marker in the proper orientation relative to either vector primer will support amplification, and the size of the PCR product indicates the distance between the vector-insert junction and the STS. In Figure 7.6, only the top subclone contains STS number 1, and this subclone is chosen to generate STS number 2 from the known sequence at the opposite end. Although the illustration shows the STS as being generated from the outermost edge of the known sequence, any convenient region of the known sequence can be used. In this example, the top five subclones all contain STS number 2, and the sizes of the PCR products allows them to be ordered according to the position of the STS. As in spectrum PCR, the subclone with the insert extending maximally to the right is chosen to continue the walk (in this example the subclone marked with two asterisks), and an STS derived from the right-hand end of the subclone becomes STS number 3. PCR with STS number 3 enables identification and ordering of the subclones containing it, and the subclone indicated with three asterisks is chosen to continue the walk. Eventually, after the entire walk is completed, the set of 96 subclones is ordered as shown in the illustration. Progress in the walk can be accelerated beyond the rate implied in Figure 7.6 because the walk need not be oriented from left to right. The walk can be conducted bidirectionally from both ends of the target sequence or there can even be multiple internal starting points since all of the subclone ends are sequenced in advance. All that is required for multiple starting points is a system of data handling sufficient to identify when the walks coalesce owing to the presence of two STS markers in a single subclone.

Assuming an 80-kb target sequence, the ends of the 96 subclones used in ordered shotgun sequencing will be separated by an average of approximately 400 bp. Since the end sequences themselves are at least 350 bp, there is little that needs to be filled in for closure. Longer reads in the beginning reduce the closure problem still further. Closure can be accomplished by sequencing off the STS primers used in the walk, longer reads off the vector ends of critical templates, or primer walking when necessary. It may even be cost effective to include a somewhat greater redundancy of shotgun sequencing at the beginning, or as the mapping proceeds, either with additional mapping clones or with shorter double-stranded or single-stranded sequencing templates. In any case, the 96 ordered subclones with sequenced ends affords an ultrahigh resolution STS map of the target sequence that is likely to be of great utility in itself (for example, for screening cDNA libraries for expressed sequences) and may well guide the choice of regions with priority for closure.

Although ordered shotgun sequencing may well have advantages over both pure strategies of shotgun or directed sequencing, it should be emphasized that the procedure is still in the testing and evaluation stages. For sequencing P1 clones, the only proven technologies are shotgun sequencing and directed sequencing off transposon insertions into 3-kb templates ordered by spectrum PCR. The apparent theoretical advantages of ordered shotgun sequencing may therefore be offset by unforeseen difficulties in subclone generation, PCR mapping, or some other component in the procedure. For *Drosophila*, the critical issue is not how the P1 clones will be sequenced but that they will be sequenced with a timetable geared to approaching completion of the sequence of the euchromatic genome by the year 2000.

REFERENCES

1. Oliver SG, and 146 other authors. The complete DNA sequence of yeast chromosome III. Nature 1992; 357:38-46.
2. Cooley L, Kelley R, Spradling A. Insertional mutagenesis of the *Drosophila* genome with single *P* elements. Science 1988; 239: 1121-1128.
3. Cooley L, Berg C, Spradling A. Controlling *P* element insertional mutagenesis. Trends Genet 1988; 4:254-258.
4. Zhang P, Spradling AC. Insertional mutagenesis of *Drosophila* heterochromatin with single *P* elements. Proc Natl Acad Sci USA 1994; 91:3539-3543.
5. Smith D, Wohlgemuth J, Calvi BR, Franklin I, Gelbart WM. *Hobo* enhancer trap mutagenesis reveals an insertion specificity different from *P* elements. Genetics 1993; 135:1063-1076.
6. Watson JJ, Cook-Deegan RM. Origins of the human genome project. FASEB J 1991; 5:8-11.
7. Collins F, Galas D. A new five-year plan for the U.S. human genome project. Science 1993; 262:43-49.
8. Deininger PL. Random subcloning of sonicated DNA: Application to shotgun DNA sequence analysis. Anal Biochem 1983; 129: 216-223.
9. Bankier AT, Weston KM, Barrell BG. Random cloning and sequencing by the M13/Dideoxynucleotide chain termination method. Methods Enzym 1987; 155:51-93.
10. Bodenteich A, Chissoe S, Wang Y-F, Roe BA. Shotgun cloning as a strategy of choice to generate templates for high-throughput dideoxynucleotide sequencing. In J. C. Venter, ed. *Automated DNA Sequencing and Analysis Techniques.* London: Academic Press 1994; 42-50.
11. Pan H-Q, Wang Y-P, Chissoe SL, Bodenteich A, Wang Z, Iyer K, Clifton SW, Crabtree JS, Roe BA. The complete nucleotide sequences of the pSacBII P1 cloning vector and three cosmid cloning vectors: pTCF, svPHEP, and LAWRIST16. Gene Analysis Tech-

niques and Applications 1995 (in press).

12. Daniels DL, Plunkett G, Burland V, Blattner FR. Analysis of the *Escherichia coli* genome: DNA sequence of the region from 84.5 to 86.5 minutes. Science 1992; 257:771-778.
13. Sulston J, Du Z, Thomas K, Wilson R, Hillier L, Staden R, Halloran N, Green P, Thierry-Mieg J, Qiu L, Dear S, Coulson A, Craxton M, Durbin R, Berks M, Metzstein M, Hawkins T, Ainscough R, Waterston R. The *C. elegans* sequencing project: a beginning. Nature 1992; 356:37-42.
14. Wilson R, Ainscough R, Anderson K, Baynes C, Berks M, Bonfield J, Burton J, Connell M, Copsley T, Cooper J, Coulson A, Craxton M, Dear S, Du Z, Durbin R, Favello A, Fraser A, Fulton L, Gardner A, Green P, Hawkins T, Hillier L, Jier M, Johnston L, Jones M, Kershaw J, Kirsten J, Laisster N, Latreille P, Lightning J, Lloyd C, Mortimer B, O'Callaghan M, Parsons J, Percy C, Rifken L, Roopra A, Saunders D, Shownkeen R, Sims M, Smaldon N, Smith A, Smith M, Sonnhammer E, Staden R, Sulston J, Thierry-Mieg J, Thomas K, Vaudin M, Vaughan K, Waterston R, Watson A, Weinstock L, Wilkinson-Sproat J, Wohldman P. 2.2 Mb of contiguous nucleotide sequence from chromosome III of *C. elegans*. Nature 1994; 368:32-38.
15. Strathmann M, Hamilton BA, Mayeda CA, Simon MI, Meyerowitz EM, Palazzolo MJ. Transposon-facilitated DNA sequencing. Proc Natl Acad Sci USA 1991; 88:1247-1250.
16. Yoshida K, Strathmann MP, Mayeda CA, Martin CH, Palazzolo MJ. A simple and efficient method for constructing high resolution physical maps. Nucleic Acids Res 1993; 21:3553-3562.
17. Barnes WM. PCR amplification of up to 35-kb DNA with high fidelity and high yield from λ bacteriophage templates. Proc Natl Acad Sci USA 1994; 91:2216-2220.

CHAPTER 8

GENOME EVOLUTION

In addition to the bacteriophage P1 library of *D. melanogaster*, we have also produced a P1 library of *D. virilis*.[1] The long-range purpose of this library is to provide the materials for a comprehensive study of the evolution of genome structure and organization. The library is also of immediate use in isolating genes in *D. virilis* that are homologs of those in *D. melanogaster* for the purpose of identifying noncoding regions that are conserved between the species and therefore likely to be important in gene regulation.

The first part of this chapter is similar to chapter 3 in providing background. The rich history of evolutionary studies in *Drosophila* is only recently emerging from a period of eclipse owing to the revival of interest in interspecific comparisons among molecular geneticists. These studies and other considerations also provide the rationale for the choice of *D. virilis* as a standard species of comparison with *D. melanogaster*.

LORD OF THE FLY ROOM

Most geneticists are well aware that the *Drosophila* genome project is the culmination of the tradition of *Drosophila* genetics begun by Thomas Hunt Morgan more than 80 years ago. By 1909 Morgan had already decided to try *Drosophila* as a model organism for genetic studies.[2] Coincidentally, 1909 was also the year that Morgan was persuaded—for the first and only time in his 24 years at Columbia—to give the opening lectures in the undergraduate course in Zoology. Among the students were Alfred H. Sturtevant and Calvin B. Bridges. Both were attracted to Morgan and asked to be taken into his laboratory.[3] Thus began the famous "fly room" at Columbia: a 16 x 23 foot room crammed with eight desks and brilliant young people pursuing a wide-open subject. Within a few years *Drosophila* was the premier organism for experiments in animal genetics. Its small size, large number of progeny, and ease of culture and manipulation made it ideal for studies of transmission genetics and mutagenesis. In the early 1930s the giant, exquisitely banded, polytene chromosomes were discovered in the larval salivary glands, and *Drosophila* soon became one of the preeminent organisms for cytogenetics.[4] Through the years the discovery of thousands of different mutations and chromosome rearrangements has multiplied and expanded the opportunities for the genetic manipulation and study of *Drosophila*. Today there are approximately 3,800 described mutations affecting virtually every aspect of the organism's morphology, development, and behavior.[5] That is why, more than 80 years after Morgan first decided to give *Drosophila* a try, it remains one of the key model organisms for research in molecular biology, cell biology, developmental biology, neurobiology, and evolution. Therefore, it was perfectly logical, and even inevitable, that *Drosophila* was included among the select few model organisms whose complex genomes were chosen for study under the auspices of the Human Genome Project.[6]

TWO TRADITIONS IN *DROSOPHILA* GENETICS

There is also a second tradition of *Drosophila* genetics that overlaps with that established by Morgan but should be distinguished from it. This tradition has become associated with the names John T. Patterson and Wilson S. Stone who carried out important studies at the University of Texas between the 1930s and the 1950s on the natural history and evolution of *Drosophila* species.[7] Both traditions of *Drosophila* genetics can trace their intellectual ancestry (at least in part) to Louis Agassiz who founded the Museum of Comparative Zoology at Harvard.[3] Ironically, Agassiz was one of the last great antidarwinians of the 19th century. The fundamental distinction between the traditions is epistemological: the Morgan school was mechanistic, the Patterson-Stone school evolutionary; the former sought to understand how genetic processes work, the latter to understand why they may differ from one species to the next. Such simple contrasts are necessarily caricatures of more complex and balanced views: the traditions of classical and evolutionary *Drosophila* genetics are not in any sense contradictory or mutually exclusive, and many geneticists have been able to cross over between the traditions and make important contributions to both—for example, Alfred H. Sturtevant, Hermann J. Mueller, and Theodosius Dobzhansky.

The *Drosophila* polytene chromosomes gave strong impetus to studies of chromosome evolution for the following reasons emphasized by Stone[8] in 1962 (p. 528): "The polytene chromosomes are the largest and most complex ordered living system with a determined and consistent inherited pattern that is changed in geometrically simple and limited ways through two-break rearrangements. No other cellular organelle in natural populations of living organisms has an equivalent ordered complexity in which changes can so easily be determined. Therefore chromosome evolution in the genus *Drosophila* and in other Diptera with polytene chromosomes is most readily and accurately measurable." In other words, the chromosome banding patterns are particularly suited to evolutionary studies because they are linear and because they are unique. In emphasizing linearity and uniqueness, Stone had seized upon precisely those features that polytene chromosomes share with DNA and protein sequences that make these macromolecules so attractive for phylogenetic analysis.[9] Indeed, the banding patterns of *Drosophila* chromosomes have the additional advantage that there are few or no homoplasies (similarities in banding patterns owing to the fixation of convergent, parallel, or reverse mutations).

REVIVAL OF INTEREST IN SPECIES COMPARISONS

With the rise of molecular biology particularly in the 1960s, the Morgan tradition came to dominate *Drosophila* genetics and the evolutionary tradition became overshadowed. For almost 20 years it was in eclipse. The evolutionary tradition continued in population genetics[10] but seemed increasingly to be irrelevant to the problems that drove mainstream molecular genetics—How are genes regulated? How do they control development? How do they organize the nervous system?

Remarkably (perhaps predictably), as the focus of molecular geneticists changed from sequence to function, there has been a revival in interspecific comparisons—the primary approach of the evolutionary tradition. The rationale of interspecific comparisons in molecular genetics is that blocks of sequence that are highly conserved in distant species are likely to be sequences critical for function. This principle is implicitly accepted and widely implemented in searches for shared amino acid motifs among proteins found in different organisms.[11] In the context of *Drosophila*, however, the comparisons are less informative for amino acid sequences than they are for the noncoding regions such as introns and the noncoding regions upstream and downstream of coding regions. In species of *Drosophila*, amino acid sequences are usually quite similar owing to the relatively slow rates of amino acid replacement. The noncoding regions are typically less constrained and so evolve at faster rates. For example, between *D. melanogaster* and *D. virilis*, the average number of synonymous nucleotide substitution per synonymous site (K_S) is approximately 1.15.[12] Assuming a Poisson distribution of substitutions, this number implies that only about 32% of synonymous nucleotides would have escaped substitution by chance alone.

A practical test of the utility of the comparative approach for identifying critical regulatory regions was carried out by Fortini and Rubin[13] who studied the in vivo expression of 31 mutations in the promoter region of an opsin gene expressed in R7 photoreceptor cells. Among these mutations, seven had detectable effects on expression of the opsin gene: all seven of these mutations were located within blocks of sequence conserved in *D. virilis*. Furthermore, only one mutation located within a conserved block had no detectable effect on expression.

The most extensive interspecific comparison of contiguous sequence thus far completed is that of approximately 160 kb in the region of the *Antennapedia* complex.[14] One small part of the total analysis included the promoter region of the Antp-P1 transcript, the first exon, and the first intron, comprising approximately 6 kb; among noncoding sequences in this region the longest conserved blocks were upstream from the first exon (119 of 127 nucleotides and 112 of 134 nucleotides) and within the first intron (38 of 43 nucleotides and 43 of 58 nucleotides). Such comparisons identify regions particularly likely to be implicated in regulation and target them for more detailed study by deletion analysis, directed mutagenesis, and other methods. Similar comparisons of genes from *D. melanogaster* with those in *D. virilis* include those of *engrailed* (actually an analysis of heteroduplex structures),[15] *dopa decarboxylase*,[16] *seven-in-absentia* and *rhodopsin-4*,[17] *hunchback*,[18] *rough*,[19] *sevenless*,[20] the 3' end of *bicoid*,[21] and *Ultrabithorax*.[22] This is an incomplete list. The GenBank release 82.0 (Spring 1994) includes more than 50 coding sequences from *D. virilis* that are homologous to sequences from *D. melanogaster*.

CHOICE OF SPECIES: RATIONALE FOR *D. VIRILIS*

Because many comparisons have already been carried out with *D. virilis*, this species has become the "default" organism for interspecific comparisons. The initial choice of *D. virilis* was not arbitrary, however. *D. virilis* is a logical choice as a standard of comparison of DNA sequences from *D. melanogaster* because of the evolutionary distance between the species. Figure 8.1 illustrates the phylogenetic relationship between the two major subgenera of *Drosophila*. The subgenus *Sophophora* includes approximately 275 named species (among them *D. melanogaster* and *D. pseudoobscura*) and the subgenus *Drosophila* includes approximately 800 named species (among them about 300 in the Hawaiian *Drosophila*). Figure 8.1 embodies some uncertainty about the most useful classification of the celebrated Hawaiian *Drosophila*. Based on an analysis of 217 adult morphological characters, Grimaldi[23] concluded that the Hawaiian species were sufficiently divergent to warrant assignment to a new genus, *Idiomya*. However, since the molecular evidence is overwhelming in support of monophyly of the Hawaiian *Drosophila* and the rest of the subgenus *Drosophila*, it has been

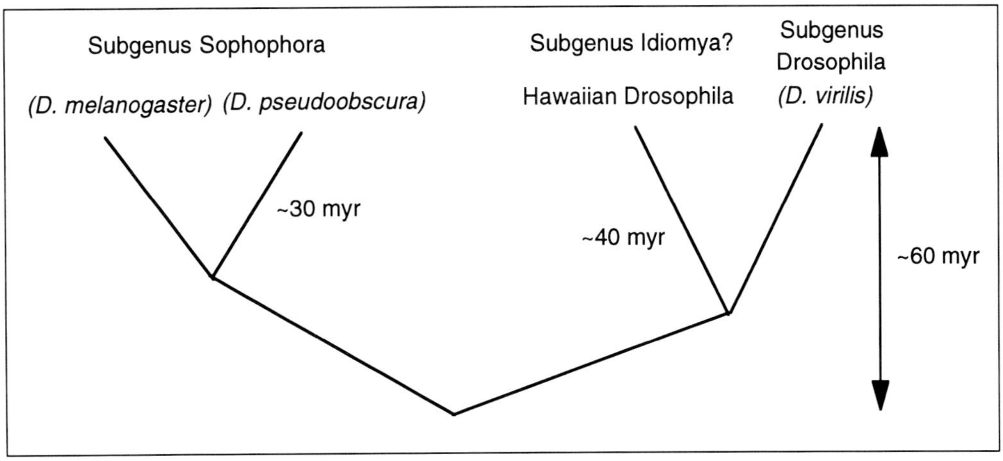

Fig. 8.1. Times of divergence within and between major groups of the subgenera Sophophora *and* Drosophila. *The species* D. melanogaster *and* D. pseudoobscura *represent the Sophophoran species groups* melanogaster *and* obscura, *respectively. The Hawaiian* Drosophila *may ultimately be placed into a separate subgenus,* Idiomya. *The divergence times are approximate and may be too long by as much as a factor of two.*

suggested that the Hawaiian *Drosophila* should be ranked at most as a new subgenus if they are to be taxonomically distinguished at all.[24]

The divergence times in Figure 8.1 are also somewhat uncertain. They are in line with the usually quoted times for these groups and are consistent with other estimates.[25] The possible errors are on the order of a factor of two and the times are more likely to be overestimated than underestimated. A recent analysis of all the pertinent data suggest a range of 30-60 million years for the *Sophophora-Drosophila* divergence, 20-40 million years for the *Idiomya-Drosophila* divergence, and 25-40 million years for the obscura (*D. pseudoobscura)*—melanogaster (*D. melanogaster*) divergence.[24,26]

As shown in Figure 8.1, *D. virilis* is a member of the subgenus *Drosophila* and so is as distantly related to *D. melanogaster* as a species can be and still be regarded as a bone fide member of the genus. With a divergence time as long as 60 million years, the nucleotide sequences that are conserved between *D. virilis* and *D. melanogaster* are likely either to be functionally important or else exceptionally lucky in avoiding chance fixation of nearly neutral mutations. As noted, the rates of silent-site substitution between *D. melanogaster* and *D. virilis* suggest that only about 32% of synonymous nucleotides would have escaped substitution by chance alone. Therefore, in comparing *D. virilis* with *D. melanogaster*, any block of six or more conserved nucleotides flanked by divergent sequences may be taken as strong suggestive evidence of selective constraint. (Nominally $P < 0.001$ not correcting for the fact that most tests include multiple comparisons.)

THE GENOME OF *D. VIRILIS*

D. virilis and its close relatives (collectively called the virilis species group) are among the species whose polytene chromosomes are exceptionally favorable for analysis because they are larger and more distinctly banded even than those of *D. melanogaster*. The DNA content of *D. virilis* is also among the largest in the genus *Drosophila*—approximately 313 Mb compared to the approximately 165 Mb of *D. melanogaster*.[27,28] Some of the larger size is attributable to a greater proportion of heterochromatin (including satellite DNA) in the *D. virilis* genome. Approximately half of the genome of *D. virilis* consists of heterochromatin as compared with approximately one-third in *D. melanogaster*.[27,29-31] The euchromatic genome of *D. virilis* is therefore about 150 Mb compared with about 110 Mb in *D. melanogaster*. What accounts for the difference of 40 Mb of DNA in such similar organisms?

Throughout the genus *Drosophila*, the chromosome arms can usually be put into correspondence with the chromosomal elements thought to be ancestral in the genus.[32] Owing to the fusion of acrocentrics or the splitting of metacentrics, the actual number of chromosomes may range from 3 to 6.[33] *D. virilis* is unexceptional in having an acrocentric X chromosome, an acrocentric Y chromosome, four acrocentric autosomes (numbered chromosomes 2-5), and a very small "dot" autosome (chromosome 6). This karyotype is regarded as ancestral in the genus.[32] The *D. virilis* chromosomes X, 2, 3, 4, 5, 6 correspond to the *D. melanogaster* chromosome arms X, 3R, 3L, 2L, 2R, and 4, respectively.[34]

The gene content of the chromosome arms is generally conserved in *D. melanogaster* and *D. virilis*,[1,17,34-36] although the full extent of chromosome-arm conservation is unclear owing to the relatively low density of genes that are known from DNA sequences (as opposed to phenotypes of mutations) to be homologous between the species. Despite the general conservation of gene content, there is considerable scrambling of the physical locations of the genes along the arms because of the fixation of paracentric inversions. Inversions are the primary source of genome rearrangements in *Drosophila* evolution. Translocations are rare in *Drosophila* but common in mammals.[37,38] The rarity of fixed translocations in *Drosophila* is probably due to the larger effective population size resulting in a smaller likelihood of chance fixation of reciprocal translocations owing to their reduced fertility.[39]

Based on the relative positions of homologous genes on the X chromosome of *D. virilis* and *D. melanogaster*, we have estimated the rate of fixation of inversion breakpoints per chromosome arm to be approximately 0.4 ± 0.4 fixations per million years.[39] Given a divergence time between *D. melanogaster* and *D. virilis* of 120 million years (60 million years on each branch), the number of fixed breakpoints per chromosome arm between the species is expected to be approximately 48—roughly one break per genetic map unit. These estimates are based on a very small number of genes, and it will be interesting to learn how they hold up with additional data.

It should also be noted that *D. virilis* has been the object of extensive classical genetic analysis using mutants: approximately 300 loci have been identified by mutation,[34,40] many of the gene homologies with *D. melanogaster* have been identified,[34] and an efficient system of hybrid dysgenesis for transposon tagging has been described.[41,42]

THE VIRILIS SPECIES GROUP

The virilis species group is also of considerable interest in connection with the evolution of reproductive isolation. The species *D. virilis* was first described by Sturtevant[43] in 1916, and at present 12 species are assigned to the virilis group. The virilis group is typically divided into two phylads of even more closely related species—the virilis phylad and the montana phylad (Fig. 8.2). The virilis phylad includes *D. virilis*, *D. lummei*, *D. novamexicana*, *D. americana*, and *D. texana*; and the montana phylad includes *D. montana*, *D. lacicola*, *D. flavomontana*, *D. borealis*, *D. littoralis*, *D. ezoana*, and *D. kanekoi*.[44] The phylogenetic relationships among

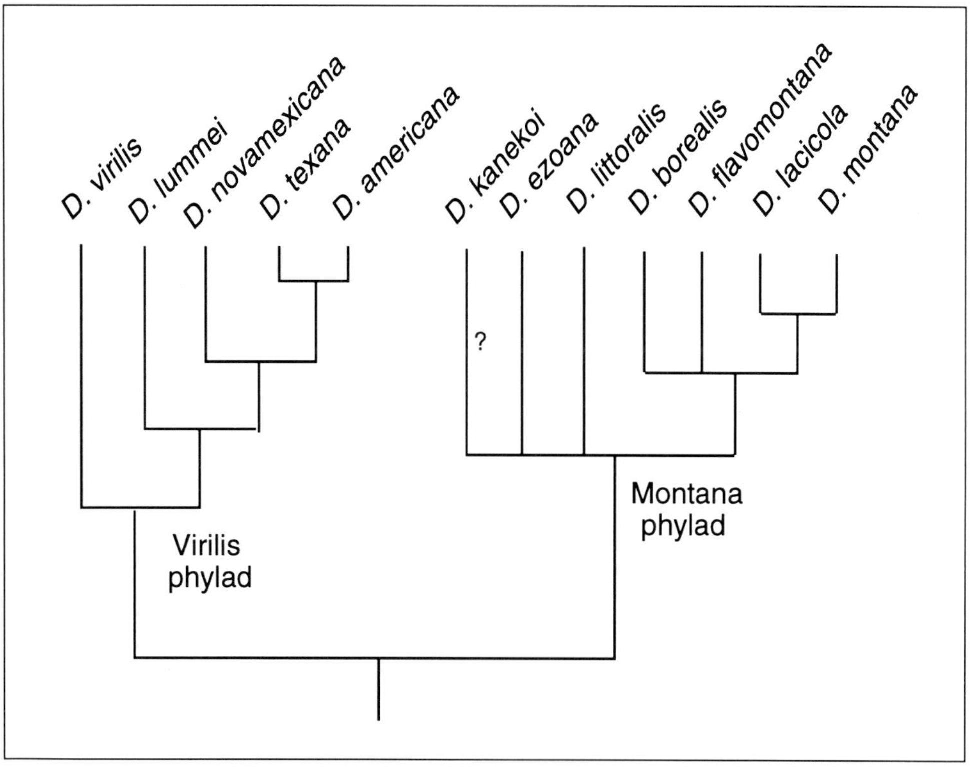

Fig. 8.2. Phylogenetic relationships among species in the virilis species group as inferred from inversions. The position of D. kanekoi *is somewhat uncertain.*

the species are quite well established from chromosomal comparisons and there are only minor ambiguities. In pairwise comparisons, the species differ in more than 100 chromosomal inversions that are either fixed or polymorphic, as well as in several chromosomal fusions and other structural variations.[7,44,45]

The species in the virilis group illustrate the various intergradations of genetic relationships, species differences, and reproductive isolation that exist in closely related species derived from a common ancestor:[7,46] among 45 interspecific crosses in the species group that have been examined, 25 yield viable and fertile or semifertile hybrids of one or both sexes, and *D. virilis* itself is interfertile with all species in the virilis phylad and with four of the seven species in the montana phylad.[44]

P1 LIBRARY OF *D. VIRILIS*

The principal limitation of the analysis of polytene banding patterns is that the differences in banding patterns among all but the most closely related species are usually too great for homologous chromosomal regions to be identified reliably. For example, there is no similarity whatsoever between the banding patterns of *D. virilis* and *D. melanogaster*, even in regions known to contain homologous genes. Cytological analysis alone therefore results in a very incomplete understanding of the patterns, processes, and functional significance of genome evolution in *Drosophila*.

The simplest approach to comparative genome mapping would be to use the mapped P1 clones from *D. melanogaster* to identify homologous regions in other species. With respect to *D. virilis*, this procedure is effective with highly conserved genes,[35] but it does not seem to be generally applicable. In our laboratory, we were unable to obtain consistent hybridization with the polytene chromosomes of *D. virilis* when using P1 clones from *D. melanogaster* as probes.[1] The simplest explanation is that, for most genes, there has been too much DNA sequence divergence in the time since *D. virilis* and *D. melanogaster* last shared a common ancestor.

In order to provide the material for a clone-based physical map of the genome of *D. virilis*, we constructed an arrayed library of 10,080 P1 clones in the vector pAd10 sacBII.[1] The distribution of insert sizes is illustrated in Figure 2.5 the average insert size is 65.8 kb with approximately 75% of the clones having inserts exceeding 50 kb.

Assuming a euchromatic genome size of 150 Mb, the clones in the *D. virilis* library include about 4.4 haploid genome equivalents.

Working toward the ultimate objective of generating a framework P1 map of *D. virilis*, a sample of 186 randomly selected clones was mapped by in situ hybridization with the salivary gland chromosomes. As far as could be determined from such a small sample, the sites of hybridization in the euchromatin appeared to be more or less randomly distributed. Interestingly, a significant fraction of these clones hybridized to multiple euchromatic sites but not to the chromocenter, which is a pattern of hybridization that is very rare among clones derived from *D. melanogaster*.

How can the framework maps of *D. virilis* and *D. melanogaster* be correlated? We propose the use of genetic landmarks derived from sequenced genes in *D. melanogaster*. Specifically, we have developed a method for identifying *D. virilis* clones containing homologs of *D. melanogaster* genes by means of colony hybridization of the P1 library with specific probes obtained from *D. melanogaster* with the polymerase chain reaction.[1] This method proved successful for 9 of 10 genes tested. Since the generation of P1 contigs in the *D. melanogaster* library will require the synthesis of oligonucleotides capable of amplifying each of approximately 700 *D. melanogaster* genes (chapter 6), PCR probes from these genes are the primary candidates for hybridization to correlate the framework maps.

GENOME EVOLUTION IN SUBGENUS *SOPHOPHORA*

It has been noted that, except for very closely related species, the polytene banding patterns are usually too dissimilar to allow homologous regions to be identified. This limitation applies even between species whose morphological similarity implies virtual certainty of close relationship.[8] Although in situ hybridization with P1 clones from *D. melanogaster* has not been generally reliable in *D. virilis* (subgenus *Drosophila*), better results have been obtained with *D. pseudoobscura* (subgenus *Sophophora*).

D. pseudoobscura is a member of the obscura species group that includes such species as *D. pseudoobscura, D. miranda, D. subobscura, D. guanche,* and *D. affinis*[47] Although the correspondences between the chromosome arms of *D. melanogaster* and various species in the obscura group have been ascertained by in situ hybridization with probes from *D. melanogaster*,[48] no detailed, point-by-point cytological comparison between the genomes has been carried out. The availability of the framework P1 map of *D. melanogaster* makes such a study feasible. In a preliminary test of 25 P1 clones from the X chromosome of *D. melanogaster*, 20 yielded strong, single hybridization patterns with the X chromosome of *D. pseudoobscura* (Segarra and Lozovskaya, unpublished observations). No ambiguities resulting from hybridization with transposable elements were found, probably because, relative to single-copy genes, the transposable elements are sufficiently divergent in sequence.[49] (Interestingly, only 12 of the 25 clones yielded satisfactory hybridization signals with *D. subobscura*). Hence, the framework map of *D. melanogaster* provides the material for a detailed alignment of the *D. melanogaster* and *D. pseudoobscura* euchromatic genomes with a density of markers of approximately one every 100 kb. Many of the 80-kb P1 clones will also contain sites of rearrangement breakpoints between the species and thereby provide the material for determining the molecular mechanisms, as well as the patterns, of genome evolution.

PROSPECTS

Part of the reason that the tradition of the Morgan school came to overshadow that of the evolutionary school was the preference for process over description and for mechanism over result. Given the inherent limitations of cytological analysis, it was inevitable that most evolutionary studies would be largely phenomenological. Being unable to recognize homologies in banding patterns except in closely related species, and being unable to carry out crosses to study genetic differences between species, what beyond phenomenology was possible?

One of the ironies of modern biology is that the great success of the Morgan school (and of molecular biology in general) has made it possible to overcome the limitations of traditional evolutionary studies. Comparative physical mapping compensates for the inability to recognize homologous regions through similarities in banding patterns. Comparative sequence analysis compensates for the inability to study genetic differences through crosses. So there has been a revival of interest in interspecific comparisons—but with a strong emphasis on mechanism rather than phenomenology. The new approach therefore warrants the term "genome evolution" to recognize its strong emphasize on mechanism. Among the important issues for genome evolution are questions like: What are the principal molecular mechanisms by which inversions are created in natural populations? Are these mechanisms the same in all species groups? What features of genome organization are conserved in *Drosophila* evolution? Why are these features conserved?

Ultimately, the enigma of genome evolution is whether variation in genome organization is a product of natural selection or whether these features of genome organization are "frozen accidents".[50] To what extent do conserved segments of the genome represent adaptively superior combinations of genes resulting from functional or regulatory interactions? To what extent are genome size, DNA composition, chromosome number, and gene organization accidents of chance events and historical contingency with no functional significance of their own?

REFERENCES

1. Lozovskaya ER, Petrov DA, Hartl DL. A combined molecular and cytogenetic approach to genome evolution in *Drosophila* using large-fragment DNA cloning. Chromosoma 1993; 102:253-266.
2. Carlson EA. *The Gene: A Critical History*. Philadelphia: W. B. Saunders 1966.
3. Sturtevant AH. *A History of Genetics*. New York: Harper & Row 1965.
4. Painter TS. A new method for the study of chromosome rearrangements and the plotting of chromosome maps. Science 1933; 78:585-586.
5. Sidén-Kiamos I, Saunders RDC, Spanos L, Majerus T, Treanear J, Savakis C, Louis C, Glover DM, Ashburner M, Kafatos FC. Towards a physical map of the *Drosophila melanogaster* genome: Mapping of cosmid clones within defined genomic divisions. Nucleic Acids Res 1990; 18:6261-6270.
6. NCHGR, DOE. Understanding Our Genetic Inheritance. The U. S. Genome Project: The First Five Years, FY 1991-1995. Washington, DC: US Department of Energy, DOE/ER-0452P 1990.
7. Patterson JT, Stone WS. *Evolution in the Genus Drosophila*. New York: Macmillan 1952.
8. Stone WS. The dominance of natural selection and the reality of superspecies (species groups) in the evolution of *Drosophila*. Univ Texas Publ 1962; 6205:507-537.
9. Hillis DM, Moritz C. *Molecular Systematics*. Sunderland MA: Sinauer 1990;
10. Lewontin RC. *The Genetic Basis of Evolutionary Change*. New York: Columbia University Press 1974.

11. Green P, Lipman D, Hillier L, Waterston R, States D, Claverie J-M. Ancient conserved regions in new gene sequences and the protein databases. Science 1993; 259:1711-1716.
12. Moriyama EN, Gojobori T. Rates of synonymous substitution and base composition of nuclear genes in *Drosophila*. Genetics 1992; 130:855-864.
13. Fortini ME, Rubin GM. Analysis of *cis*-acting requirements of the *Rh3* and *Rh4* genes reveals a bipartite organization to rhodopsin promoters in *Drosophila* melanogaster. Genes Develop 1990; 4:444-463.
14. Hooper JE, Pérez-Alonzo M, Bermingham JR, Prout M, Rocklein BA, Wagenbach M, Edstrom J-E, de Frutos R, Scott MP. Comparative studies of *Drosophila Antennapedia* genes. Genetics 1992; 132:453-469.
15. II JA, Poole SJ, Wright DK, O'Farrell PH. Electron microscope heteroduplex mapping identifies regions of the engrailed locus that are conserved between *Drosophila melanogaster* and *Drosophila virilis*. Mol Cell Biol 1985; 5:3600-3609.
16. Bray SJ, Hirsh J. The *Drosophila virilis* dopa decarboxylase gene is developmentally regulated when integrated into *Drosophila melanogaster*. EMBO J 1986; 5:2305-2311.
17. Neufeld TP, Carthew RW, Rubin GM. Evolution of gene position: Chromosomal arrangement and sequence comparison of the *Drosophila melanogaster* and *Drosophila virilis sina* and *Rh4* genes. Proc Natl Acad Sci USA 1988; 88:10203-10207.
18. Treier M, Pfeifle C, Tautz D. Comparison of the gap segmentation gene *hunchback* between *Drosophila melanogaster* and *Drosophila virilis* reveals novel modes of evolutionary change. EMBO J 1989; 8:1517-1525.
19. Heberlein U, Rubin GM. Structural and functional comparisons of the *Drosophila virilis* and *Drosophila melanogaster* rough genes. Proc Natl Acad Sci USA 1990; 87:5916-5920.
20. Michael WM, Bowtell DDL, Rubin GM. Comparison of the *sevenless* genes of *Drosophila virilis* and *Drosophila melanogaster*. Proc Natl Acad Sci USA 1990; 87:5351-5353.
21. Luk SK, Kilpatrick M, Kerr K, Macdonald PM. Components acting in localization of *bicoid* mRNA are conserved among *Drosophila* species. Genetics 1994; 137:521-530.
22. Bomze HM, López AJ. Evolutionary conservation of the structure and expression of alternatively spliced *Ultrabithorax* isoforms from *Drosophila*. Genetics 1994; 136:965-977.
23. Grimaldi DA. A phylogenetic, revised classification of genera in the *Drosophilidae* (Diptera). Bull Amer Mus Nat Hist 1990; 197:1-139.
24. Powell JR, DeSalle R. *Drosophila* molecular phylogenies and their uses. Evol Biol 1995; 28: (in press).

25. Beverley SM, Wilson AC. Molecular evolution in *Drosophila* and the higher diptera II. Time scale for fly evolution. J Mol Evol 1984; 21:1-13.
26. Kwiatowski J, Skarecky D, Bailey K, Ayala FJ. Phylogeny of *Drosophila* and related genera inferred from the nucleotide sequence of the Cu,Zn *Sod* gene. J Mol Evol 1994; 38:443-454.
27. Laird CD. DNA of *Drosophila* chromosomes. Ann Rev Genet 1973; 7:177-204.
28. John B, Miklos GLG. *The Eukaryotic Genome in Development and Evolution.* London: Allen and Unwin 1988.
29. Gall JG. Repetitive DNA in *Drosophila.* In: B. A. Hamkalo, J. Papaconstantinou, eds. *Molecular Cytogenetics.* New York: Plenum 1973; 59-73.
30. Miklos GLG. Localized highly repetitive DNA sequences in vertebrate and invertebrate genomes. In: R. J. MacIntyre, ed. *Molecular Evolutionary Genetics.* New York: Plenum 1985; 241-321.
31. Lohe AR, Brutlag DL. Identical satellite DNA sequences in sibling species of *Drosophila.* J Mol Biol 1987; 194:161-170.
32. Muller HJ. Bearings of the *Drosophila* work on systematics. In: J. Huxley, ed. *New Systematics.* Oxford, England: Clarendon 1940; 185-268.
33. Ashburner M. *Drosophila: A Laboratory Handbook.* Cold Spring Harbor, NY: Cold Spring Harbor Laboratory Press 1989.
34. Alexander ML. The genetics of *Drosophila virilis.* In: M. Ashburner, E. Novitski, eds. *The Genetics and Biology of Drosophila,* Volume 1c. New York: Academic Press 1976; 1365-1427.
35. Loukas M, Kafatos FC. The actin loci in the genus *Drosophila:* Establishment of chromosomal homologies among distantly related species by in situ hybridization. Chromosoma 1986; 94:297-308.
36. Whiting JH Jr., Pliley MD, Farmer JL, Jeffery DE. *In situ* hybridization analysis of chromosomal homologies in *Drosophila melanogaster* and *Drosophila virilis.* Genetics 1989; 122:99-109.
37. O'Brien SJ, Seuánez HN. Mammalian genome organization: An evolutionary view. Ann Rev Genet 1988; 22:323-351.
38. O'Brien SJ. Mammalian genome mapping: lessons and prospects. Current Opinion Genet Dev 1991; 1:105-111.
39. Hartl DL, Lozovskaya ER. Genome evolution: Between the nucleosome and the chromosome. In: B. Schierwater, B. Streit, G. P. Wagner, R. DeSalle, eds. *Molecular Approaches to Ecology and Evolution.* Basel: Birkhäuser Verlag 1994; 579-592.
40. Gubenko IS, Evgen'ev MB. Cytological and linkage maps of *Drosophila virilis* chromosomes. Genetica 1984; 65:127-139.
41. Lozovskaya ER, Scheinker VS, Evgen'ev MB. A hybrid dysgenesis syndrome in *Drosophila virilis.* Genetics 1990; 126:619-623.
42. Scheinker V, Lozovskaya E, Bishop JG, Corces VG, Evgen'ev MB. A long terminal repeat-containing retrotransposon is mobilized

during hybrid dysgenesis in *D. virilis*. Proc Natl Acad Sci USA 1990; 87:9615-9619.

43. Sturtevant AH. Notes on North American *Drosophilidae* with descriptions of twenty-three new species. Ann Ent Soc Amer 1916; 9:323-343.
44. Throckmorton LH. The *virilis* species group. In: M. Ashburner, H. L. Carson, J. N. J. Thompson, eds. *The Genetics and Biology of Drosophila*. New York: Academic Press 1982; 227-296.
45. Patterson JT, Stone WS, Griffen AB. The virilis complex in *Drosophila*. Genetics 1940; 25:131.
46. Orr HA, Coyne JA. The genetics of postzygotic isolation in the *Drosophila virilis* group. Genetics 1989; 121:527-537.
47. Lakovarra S, Saura A. Evolution and speciation in the *Drosophila obscura* group. In: M. Ashburner, H. L. Carson, J. N. Thompson Jr., eds. *The Genetics and Biology of Drosophila*, Volume 3b. New York: Academic Press 1982; 1-59.
48. Segarra C, Aguadé M. Molecular organization of the X chromosome in different species of the obscura group of *Drosophila*. Genetics 1992; 130:513-521.
49. de Frutos R, Peterson KR, Kidwell MG. Distribution of *Drosophila melanogaster* transposable element sequences in species of the obscura group. Chromosoma 1992; 101:293-300.
50. Ohno S. Ancient linkage groups and frozen accidents. Nature 1973; 244:259-262.

APPENDIX 1

PROTOCOLS

This Appendix includes three protocols for manipulating P1 clones that are in routine use in our laboratory:

- Preparations of DNA from 10 ml cultures of *E. coli* cells containing P1 clones (minipreps).
- Hybridization in situ with polytene chromosomes.
- Amplification of the insert-vector junctions of P1 clones by means of the polymerase chain reaction.

For ease of use, all the protocols are given in "cookbook" fashion rather than in the cryptic style that has become standard in journal articles. Protocols for the use of YAC clones are not included, in part because of space limitations and in part because most *Drosophila* researchers seem to prefer P1 clones to YAC clones for technical reasons. Readers interested in protocols for *Drosophila* YACs may consult Garza et al,[1,2] Ajioka et al,[3] Danilevskaya et al,[4] and Cai et al.[5] More general YAC protocols are compiled in Brownstein and Nelson.[6]

MINIPREPS FROM P1 CLONES

This protocol is a modification of the procedure described by Birnboim and Doly.[7] Recipes for the main ingredients (shown in **boldface**) are given at the end. For most P1 clones, the protocol yields a sufficient concentration of DNA that 15-20 µl can be used per restriction digest in a Southern blot. We typically use 1.5 µl per slide for in situ hybridizations with polytene chromosomes.

Step 1. Inoculate a bacterial culture containing the P1 clone of interest into 1 ml of **LB medium** containing 25 mg/ml of kanamycin. Grow the culture overnight with aeration at 37°C.

Step 2. Add 100 µl of the overnight culture to 10 ml of LB medium containing 25 µg/ml of kanamycin and incubate with aeration at 37° C to an optical density of OD_{550} = 0.05 to 0.1. Growth to this density typically takes from 1 to 1.5 hours. Hint: To allow thorough aeration in a shaking air or water bath, the culture volume should not exceed 30% of the volume of the container.

Step 3. Add IPTG (β-D-thiogalactopyranoside) to a final concentration of 1 mM and continue incubating the culture to an OD_{550} of 0.5 to 1.5. This incubation typically lasts about 2 hours.

(OPTIONAL: The P1 clone may also be inoculated directly into 10 ml of LB medium containing 25 µg/ml kanamycin plus 1 mM IPTG and grown at 37° C overnight with aeration. This approach requires less fussing than the procedure in Steps 1 through 3. The problem is that it does not work well with all P1 clones.)

Step 4. Spin the culture at 4,000 rpm for 10 min in a refrigerated centrifuge (4° C). Discard the supernatant.

Step 5. Add 200 µl of **Solution 1** to the pellet and resuspend. The pellet should be completely dispersed in the liquid. Transfer the resulting suspension to a microfuge tube.

Step 6. Add 300 µl of **NaOH-SDS solution** and mix well. Avoid vigorous shaking while mixing. Incubate in ice until the suspension clarifies—about 5 minutes.

Step 7. Add 300 µl of **3 M KAc** (pH 4.8) and mix gently until a whitish precipitate forms. Then shake and incubate in ice for a further 10 minutes.

Step 8. Spin at 14,000 rpm for 5 minutes in a microfuge. Transfer the supernatant to a clean microfuge tube.

(OPTIONAL: For a quick-and-dirty miniprep, at this stage add 0.6 volumes (about 0.5 ml) of isopropanol, mix, and spin immediately at 14,000 rpm for 5 minutes at room temperature. Remove the supernatant, add 1 ml of 70% ethanol, shake, centrifuge, and again discard the supernatant. Then dry the pellet and resuspend in 50 ml of **TE buffer**.)

Step 9. Add 3 µl of DNAase-free pancreatic **RNase** (10 mg/ml). Incubate for 30 minutes at 37° C.

Step 10. Add an equal volume of chloroform and vortex for 1 minute. Spin at 14,000 rpm for 5 minutes and transfer the aqueous layer to the clean microfuge tube.

Step 11. Add an equal volume of isopropanol, mix well, and centrifuge immediately at 14,000 rpm for 10 minutes at room temperature. Discard the supernatant.

Step 12. Wash the pellet by adding 1 ml of cold 70% ethanol. Shake, centrifuge again, and discard the ethanol. Dry the pellet and resuspend it in 40 µl of TE.

(OPTIONAL: For additional purification to obtain higher quality DNA preparations, add 10 ml of 4M NaCl to the solution resulting from Step 12. Mix and add 50 ml of 13% polyethylene glycol (PEG-8000). Mix again and incubate on ice for 1 hour. Centrifuge at 14,000 rpm for 15 minutes at 4° C (we carry out this centrifugation in a cold room). Remove the liquid. WARNING: at this stage the pellet is transparent, so be careful! Add 1ml of 70% ethanol, shake, spin again, remove the liquid and resuspend the pellet in 50 µl of TE buffer.

ANOTHER OPTION: Quite good DNA preparations can also be made from the solution in Step 12 as follows. Add 60 µl of 5 M ammonium acetate and mix to bring to total volume to 100 µl. Add 160 µl of ethanol, mix, incubate at room temperature for 5 minutes, and centrifuge at 14,000 rpm for 10 minutes. Remove the supernatant, add 1 ml of 70% ethanol, shake, centrifuge again, decant the supernatant, and resuspend the pellet in 50 µl of TE buffer.

Recipes for Minipreps

LB medium:	Tryptone	10 g
	Yeast extract	5 g
	NaCl	5 g

add dH_2O to bring volume to 1 l. (The pH should be in the range of pH 6 to pH 7.)

Solution 1: 50 mM glucose, 25 mM Tris-HCl (pH 8.0), 10 mM EDTA, store at 4° C.

NaOH-SDS: 0.2 N NaOH (800 mg per 100 ml), 1% SDS (10 ml of a stock solution of 10% SDS per 100 ml). Prepare weekly and store at room temperature.

3 M KAc:	Potassium acetate, 5 M stock solution	60.0 ml
	Glacial acetic acid	11.5 ml
	dH_2O	28.5 ml

Store at room temperature.

TE buffer: 50 mM Tris-HCl (pH 8.0), 5 mM EDTA.

RNAse A: For a stock solution, dissolve RNAse A in TE buffer at a concentration of 10 mg/ml. Heat at 100° C for 15 minutes, dispense in aliquots, and store at -20° C.

Hybridization in situ with Polytene Salivary-Gland Chromosomes

In the final analysis, the quality of an in situ preparation is only as good as the quality of the chromosome squash. It usually takes some time to learn how to obtain preparations that are optimally squashed and stretched.

Growth of Larvae for In Situ Hybridization

To grow the larvae, place approximately 50 to 70 adult flies into a culture bottle with instant *Drosophila* medium (Carolina Biological Supply) for about 48 hours. (We use plastic urinalysis bottles for such cultures.) Remove the adult flies and store the bottle at room temperature for another 24 hours. The food should remain moist during this period. When the larvae start hatching, pour a layer of a suspension of live baker's yeast over the top of the instant medium and transfer the bottle to 16 to 18° C. Larvae of the third instar will be ready for preparation of slides in 6 to 8 days.

SUBBING SLIDES (OPTIONAL)

Step 1. Load slide racks with slides. A convenient number is 60 (two 30-slide racks).

Step 2. Immerse slide racks in **subbing solution.**

Step 3. Remove racks from subbing solution, allow excess solution to run off, blow off large drops that remain with an air hose, and dry in a fume hood.

Step 4. Place racks in **SSC/Denhardt's** in a 65° C oven for 1-4 hours.

Step 5. Place slide racks in 4-liter beakers and rinse with dH_2O.

Step 6. Place racks in **ethanol/acetic acid** for 20 minutes.

Step 7. Dry in a fume hood. When completely dry, return slides to original boxes.

RECIPES FOR SUBBING SLIDES

These solutions have a shelf life of about 5 days when stored at 4° C.

Subbing solution: 0.1% gelatin, 0.01% $CrK(SO_4)_2 \cdot 12\ H_2O$. To prepare, dissolve 1.0 g of gelatin in 1 liter of hot dH_2O. After the solution has cooled, add 100 mg of chromium potassium sulfate. Use at room temperature.

SSC/Denhardt's: In this recipe, 20 × SSC is a solution of 3 M NaCl and 0.3 M sodium citrate. To prepare 1 liter of SSC/Denhardt's, mix: 150 ml of 20 × SSC, 200 mg bovine serum albumin (BSA), 200 mg polyvinylpyrolidone, 200 mg Ficoll, and 850 ml dH_2O. Use at 65° C.

Ethanol/acetic acid: For 1 liter, mix 236 ml concentrated acetic acid with 764 ml 95% ethanol. Use at room temperature.

POLYTENE CHROMOSOME PREPARATION

This procedure is modified from Atherton and Gall.[8] Properly prepared slides may be stored at room temperature for as long as 3 or 4 months before hybridization.

Step 1. Remove a larva from a bottle and place it in a large drop of 45% acetic acid.

Step 2. Carefully dissect the salivary glands and clean away the fat body from each. (It is very important to remove as much of the fat bodies as possible.)

Step 3. Transfer the dissected glands to a fresh drop of 45% acetic acid for 1-2 minutes, then transfer into 10 µl of the **1 : 2 : 3 solution** on a siliconized coverslip (size 22 mm^2).

Step 4. Cover the salivary glands with a subbed slide, spread the chromosomes by tapping the coverslip with the rubber eraser on a wood pencil until desirable spreading occurs. The degree of spreading can be determined by checking the slide under a phase-contrast microscope. After tapping, cover the slide with folded paper towel and press *very* hard to remove excess liquid and to flatten the chromosomes between the slide and coverslip.

Step 5. Immerse the slide in liquid nitrogen for 30 seconds (until the boiling stops) and, using a single-edge razor blade *carefully*, flip off the coverslip. Transfer the slide to 95% ethanol, soak for 10 minutes, and dry in air.

Recipes for Chromosome Preparation

1 : 2 : 3 solution: Mix volume-to-volume, 1 part acetic acid : 2 parts dH_2O : 3 parts glacial acetic acid.

Treatment of Slides before hybridization

Place slides containing squashed chromosomes in slide racks and carry out the following series of treatments by immersion in each solution in turn. Use a fresh solution for each of the washes done twice.

2 × SSC	65° C	30 minutes
70% ethanol	room temperature	1 minute
70% ethanol	room temperature	1 minute
95% ethanol	room temperature	2 minutes
Dry in air, then		
0.07 N NaOH	room temperature	2.5 minutes
2 × SSC	room temperature	5 minutes
2 × SSC	room temperature	5 minutes

70% ethanol	room temperature	1 minute
70% ethanol	room temperature	1 minute
95% ethanol	room temperature	2 minutes

Dry in air

DNA LABELING

This procedure is modified from Feinberg and Vogelstein.[9] We use DNA prepared by the alkaline lysis miniprep protocol described above. The labeling procedure supplies sufficient DNA for hybridization with one slide.

Step 1. Remove an aliquot of 1-5 µl of DNA solution (this volume should contain 50-100 ng of DNA) and add dH_2O to bring the total volume to 9 µl.

Step 2. Boil for 5 minutes in a water bath and place immediately in ice.

Step 3. Add 6 µl of **labeling cocktail.** Mix well and store overnight at room temperature.

Step 4. Add 3 µl of 5 M NaCl

Step 5. Add 5 µl of a 10 mg/ml solution of single stranded salmon-sperm DNA

Step 6. Add 35 µl of 95% ethanol and incubate at -20° C for 1 hour.

Step 7. Centrifuge for 20 minutes, decant the supernatant, wash pellet with 75% ethanol, centrifuge, decant the supernatant, dry the pellet, and redissolve in 12 µl of dH_2O.

Step 8. Boil in a water bath for 5 minutes and immediately place in ice.

Step 9. Add **hybridization mix.**

RECIPES FOR DNA LABELING

Hybridization mix: Combine 2.1 µl of **20 × SSC,** 4.2 µl of 50% dextran sulfate, 10.5 µl of deionized formamide, and 1.8 µl of a 10 mg/ml solution of single stranded salmon-sperm DNA.

20 × SSC contains 3 M NaCl, 0.3 M sodium citrate.

Labeling cocktail: Mix 0.6 µl of a 10 mg/ml solution of bovine serum albumin (BSA), 3.75 µl of **4 × Hexamer buffer for dATP,**

1 µl biotin-dATP (as provided by manufacturer), 0.2 µl of a solution of 5 U/µl Klenow DNA polymerase, and 0.45 µl dH_2O.

4 × **Hexamer buffer for dATP:** The following recipe makes a total volume of 2320 µl. Final concentrations are shown at the left, amounts to aliquot from stock solutions are shown at the right.

200 mM Tris-HCl	464 µl of 1M stock, pH 8
20 mM $MgCl_2$	46.4 µl of 1M stock
20 mM DTT	46.4 µl of 1M
80 µM dCTP	1.86 µl of 100 mM stock
80 µM dGTP	1.86 µl of 100 mM stock
80 µM dTTP	1.86 µl of 100 mM stock
800 mm Hepes	928 µl of 2 M stock, pH 6.6
50 U Hexamer	500 µl of 100 U/ml stock solution
dH_2O	329.6 µl

Note: If you prefer to use a biotin-labeled dNTP other than dATP, you must change the set of dNTPs included in the buffer.

Hybridization and Signal Detection

Step 1. Place the hybridization mix containing the labeled DNA on top of the chromosome squash and cover with a siliconized coverslip (20 mm^2). Place the cover slip carefully to minimize the formation of bubbles and tease out any that may have formed.

Step 2. Seal with rubber cement and incubate overnight at 37° C in a container humidified with wet paper towels.

Step 3. Remove the rubber cement and place the slide in **2 × SSC** at 37° C in order to loosen the coverslip.

Step 4. Remove the coverslip and wash as follows:

2 × SSC	37° C	10 minutes
2 × SSC	37° C	10 minutes
2 × SSC	room temperature	15 minutes
1 × PBS	room temperature	5 minutes

Step 5. Add 22.5 µl of **Complex** on top of the chromosome squash, cover with a coverslip (22 mm × 40 mm) and tease out any bubbles.

Step 6. Place the slide in a closed container humidified with wet paper towels and incubate at 37° C for 50 minutes.

Step 7. Immerse in 1 × PBS at room temperature briefly in order to loosen the coverslip. Remove the coverslip.

Step 8. Wash as follows:

1 × PBS	room temperature	10 minutes
1 × PBS	room temperature	10 minutes

Step 9. Immerse in **DAB** at room temperature for 40 minutes *in the dark.*

Step 10. Wash in 1 × PBS at room temperature for 1 minute.

Step 11. Counterstain the chromosomes with **Giemsa** at room temperature for 5 minutes.

Step 12. Wash briefly in dH_2O and dry in air.

Step 13. Place a drop of Permount over chromosome squash and cover with a coverslip. Tease out all bubbles. Examine chromosomes under a microscope but be careful: it takes about 24 hours for the Permount to polymerize.

RECIPES FOR HYBRIDIZATION AND SIGNAL DETECTION

Complex: The DETEK *Hrp* signal generating system for biotin detection comes from ENZO Diagnostics, Inc. (catalog number 43820). The 10 × dilution buffer for the Complex consists of 0.1 M sodium phosphate (pH 7.2), 1.5 M NaCl, 0.5% Triton-X 100, 50 mM EDTA. For one slide, combine 2.25 µl of 10 × dilution buffer, 20 µl dH_2O, and 0.25 µl DETEK *Hrp* Complex. Place the mixture over the chromosome squash.

DAB:		
	3,3'-diaminobenzidine	200 mg
	1 × PBS	350 ml
	H_2O_2	800 µl (add immediately before use)

Giemsa: To prepare a stock solution, mix 1.5 g Giemsa powder in 100 ml glycerol. Incubate at 60° C for 1 hour and add 100 ml of methanol. For staining, mix 7 ml of the stock solution with 343 ml of 1 × PBS.

10 × PBS:		
	NaCl	148 g
	Na_2HPO_4 (anhydrous)	19.9 g
	$NaH_2PO_4 \cdot 2H_2O$	9.36 g
	dH_2O	to bring the volume to 2 liters

AMPLIFICATION OF THE ENDS OF DNA FRAGMENTS CLONED IN P1

The procedure below gives the details of the method outlined in Nurminsky and Hartl.[10] Oligonucleotide primer sequences are given below for amplification of either end of the insert in vector pAd10 sacBII or for the left end of the insert in the vector pAd10.

PREPARATION OF TEMPLATES FOR PCR

Step 1. Grow cells of *E. coli* containing P1 clones in 5 ml LB broth with 1 mM IPTG (isopropyl β-D-thiogalactopyranoside) and 25 μg/ml kanamycin at 37° C overnight with aeration. Extract P1 DNA from the cells. The rapid alkaline lysis procedure for minipreparations described in the Applied Biosystems manual for the DyeDeoxy Terminator sequencing kit provides consistent results. The PEG precipitation step in this procedure may be omitted, but the RNAse treatment should be included.

Step 2. Digest the DNA with restriction enzymes in appropriate buffer. Enzymes *Msp*I, *HinP*I, *BsaH*I, and *Cla*I are suitable for use with the adapters described below. All of these enzymes are active in NE Buffer #4 (New England Biolabs). Use approximately 5% (approximately 2 μl) of the DNA extracted from each 5 ml culture and 1-2 U of enzyme (0.2 μl of a 10 U/ml solution) and bring the final volume to 10 μl with dH_2O. Incubate at 37° C for 1-2 hours. When numerous preparations are processed simultaneously, it is convenient to carry out these reactions in V-bottom microtiter plates.

Step 3. Prepare annealed adapter solution as follows:

Ingredient	*Stock solution*	*Volume*
L-bubble/RP1	200 nmole/ml	12 μl
S-bubble	200 nmole/ml	12 μl
NaCl	5 M	1 μl

Denature at 96° C for 2 minutes, cool rapidly to 40° C, cool at 1° C per second to 20° C, heat to 27° C, and incubate for 1.5 hours. Dilute in 100 μl dH_2O.

Step 4. Combine 1.5 μl of the restriction digest with 2.5 μl of **adapter ligation mixture.** (Note that this mixture includes an aliquot of the annealed adapter solution prepared in Step 3.) Incubate at 16° C for 2-3 hours or at 10° C overnight. When incubated for the longer time, overlay with a drop of mineral oil.

Step 5. Split the reaction volume in half (2 μl each) for separate amplification of each end of the insert. To each tube (or well, if using a microtiter plate) add 50 μl of the PCR cocktail containing 1 nm of the oligonucleotide *PCR-bubble* and 1 nm of either *pAd10SacBII/T7* or *pAd10SacBII/SP6* or *pAd10/Left*. The conditions for PCR amplification are

a. 5 minutes at 95° C
b. 30 seconds at 94° C
c. 1 minute at 53° C
d. 2 minutes at 72° C
e. Go to step b (repeat 24 times)
f. Hold at 4° C

Step 6. Analyze 5 μl of each reaction by electrophoresis in 1% agarose.

Step 7. Isolate products larger than 250 bp for sequencing. The PCR products are purified with Quiaquick spin columns (Quiagen Corporation). The products are suitable for sequencing with the Applied Biosystems DyePrimer M13RP1 sequencing kit.

ADAPTER OLIGONUCLEOTIDES

The adapter oligonucleotides are not phosphorylated, which allows them to be added in molar excess to drive the ligation.

L-bubble/RP1

5'-ACTGCTCGTACCACAATGGCTCAGGAAACAGCTATGACCGTCCTCTCCT-3'

S-bubble

5'-CGAGGAGAGGACTCA-3'

Oligonucleotide Primers for PCR

PCR bubble primer
5'-ACTGCTCGTACCACAATGGCT-3'
pAd10SacBII/T7
5'-GGCCGCTAATACGACTCACTA-3'
pAd10SacBII/SP6
5'-TAATTGGCCGTCGACATTTAG-3'
pAd10/Left
5'-GCGATCATGGCGACCACACC-3'

Recipes for Bubble PCR

Annealed adapter solution: See above Step 3 in the protocol.

Adapter ligation mixture: A 25 μl amount (sufficient for 10 reactions) contains

10 × ligation buffer (New England Biolabs)	2.5 μl
Annealed adapter solution	2 μl
T4 DNA ligase (1 Weiss U/ml)	3 μl
dH_2O	17.5 μl

PCR cocktail: The following recipe makes a total volume of 500 μl, which is enough for 10 PCR reactions.

Ingredient	**Stock solution**	**Volume**	**Final concentration**
PCR buffer (Promega)	10 ×	50 μl	1 ×
$MgCl_2$	1 M	0.75 μl	1.5 mM
dNTP	100 mM (25 mM each)	5 μl	1 mM
formamide	100%	12.5 μl	2.5%
PCR bubble primer	20 nmole/ml	25 μl	1 μM
pAd10SacBII/T7	20 nmole/ml	25 μl	1 μM
or pAd10SacBII/SP6	20 nmole/ml	25 μl	1 μM
or pAd10/Left	20 nmole/ml	25 μl	1 μM
Taq polymerase	5 U/μl	1 μl	0.01 U/μl
dH_2O	—	380.75 μl	

REFERENCES

1. Garza D, Ajioka JW, Burke DT, Hartl DL. Mapping the *Drosophila* genome with yeast artificial chromosomes. Science 1989; 246:641-646.
2. Garza D, Ajioka JW, Carulli JP, Jones RW, Johnson DH, Hartl DL. Physical mapping of complex genomes. Nature 1989; 340:577-578.
3. Ajioka JW, Smoller DA, Jones RW, Carulli JP, Vellek AEC, Garza D, Link AJ, Duncan IW, Hartl DL. *Drosophila* genome project: One-hit coverage in yeast artificial chromosomes. Chromosoma 1991; 100:495-509.
4. Danilevskaya ON, Kurenova EV, Pavlova MN, Link AJ, Koga A, Vellek A, Hartl DL. He-T family DNA sequences in the Y chromosome of *Drosophila melanogaster* share homology with the X-linked *Stellate* genes. Chromosoma 1991; 100:118-124.
5. Cai H, Kiefel P, Yee J, Duncan I. A yeast artificial chromosome clone map of the Drosophila genome. Genetics 1994; 136: 1385-1401.
6. Brownstein BH, Nelson DL eds. YAC Libraries: A User's Guide. New York: W. H. Freeman 1994.
7. Birnboim HC, Doly J. A rapid alkaline extraction procedure for screening recombinant plasmid DNA. Nucleic Acids Res 1979; 7:1513-1523.
8. Atherton D, Gall J. Salivary gland squashes for in situ nucleic acid hybridization studies. Dros Infor Ser 1972; 49:131-133.
9. Feinberg AP, Vogelstein B. A technique for radiolabeling DNA restriction endonuclease fragments to high specific activity: Addendum. Analyt Biochem 1984; 137:266-267.
10. Nurminsky DI, Hartl DL. Amplification of the ends of DNA fragments cloned in bacteriophage P1. Biotechniques 1993; 15:201-208.

APPENDIX 2

YAC LOCALIZATIONS

The tabulation here includes the estimated extent of euchromatic bands covered by YAC clones of the types described in chapter 2. For further information, see the references below or establish contact electronically with ftp.bio.indiana.edu in the directory flybase or flybase@morgan.harvard.edu. The clone designations are as follows:

- DY clones are in the vector pYACP-1.
- N clones are in the vector pYAC5.
- E, R, RX, and Rt clones are in the vector pYAC4.
- T clones are in the "half-YAC" vector described in Riethman HC, Moyzis RK, Meyne J, Burke DT, Olson MV. Cloning human telomeric DNA fragments into *Saccharomyces cerevisiae* using a yeast-artificial-chromosome vector. Proc Natl Acad Sci.USA 1989; 86:6240-6244.

All YAC clones are available on request from:

Ian W. Duncan, Department of Biology, Washington University, 1 Brookings Drive, St. Louis, MO 63130-4862
Tel: (314) 935-6719, Fax: (314) 935-4432

YAC clones prefixed DY, N, or E are are available on request from:

Daniel L. Hartl, Harvard University, Department of Organismic & Evolutionary Biology, 16 Divinity Avenue, Cambridge, MA 02138
Tel: (617) 496-3917, Fax: (617) 496-5854

REFERENCES

Ajioka JW, Smoller DA, Jones RW, Carulli JP, Vellek AEC, Garza D, Link AJ, Duncan IW, Hartl DL. *Drosophila* genome project: One-hit coverage in yeast artificial chromosomes. Chromosoma 1991; 100:495-509.

Cai H, Kiefel P, Yee J, Duncan I. A yeast artificial chromosome clone map of the Drosophila genome. Genetics 1994; 136:1385-1401.

Garza D, Ajioka JW, Burke DT, Hartl DL. Mapping the *Drosophila* genome with yeast artificial chromosomes. Science 1989; 246: 641-646.

YAC LOCALIZATIONS

YAC	START READ	END READ
R19-83	1A1	1B2-3
N13-23	1B1	1B10-11
E00-52	1B1	1B3
DY174	1B3	1B10
R14-34	1B8-9	1C2-4
N11-18	1B14-C1	1C3-4
N24-78	1C1	1C3
N03-85	1C2	1C3
N24-25	1D2-3	1E3-4
N01-64	1D4	1E1,2
Rt04-67	1E1	1E4-5
N04-96	1E3	1E4
N01-61	1E3	1F2
DY167	1E3	1F4
N18-90	1E3-4	1F1-2
DY723	1F4	2B1-2
N10-31	2A3-B1	2B7-8
R13-59	2B1	2B10
N01-69	2B2	2C1
DY645	2B3-4	2B12-16
N23-03B	2B15-17	2C1-2
N17-29	2E3-F1	3A2-3
N26-52	2F1	3A2
N15-58	2F4	3A1-2
DY774	3A3	3A5
R22-11	3A3-4	3C2-4
N12-54	3A8	3B6
N13-61	3B5-6	3C1-2
DY430	3C1	3C3
R07-58	3C1-2	3C5-7
R01-60	3C4-5	3C10-12
N01-51	3D1	3D4
N11-88	3D1-2	3E2-3
E03-14	3D4	3D+7
N09-07	3E1	3E2
R07-40	3F1-2	4A2-3
N16-77	3F4,5	4A2
N12-95	3F5-6	4A3-5
DY354	4B1	4B4
N21-40	4B1	4C4-5
N02-77	4C7	4C8
R17-02	4D1	4D6-7
R21-74	4D1	4D6-7
R17-24	4D1	4E1
R13-74	4D1	4F5,6
DY494	4D1-2	4E1-2
DY623	4D3-4	4E2-3
N24-66	4D3-5	4E2-3
N25-78	4D5-7	4E2-3
N18-43	4E1	4F5,6
R22-68	4E2	4F3-4
E01-30	4F5,6	5A2
N26-03B	4F9	5A2-3
DY1	5A5	5B2
DY4	5B1	5B6,7
E02-86	5C1	5C2
N04-92	5C4	5D2
N11-80	5D3-5	5E4-6
N03-70	5D5	5E1,2
N03-33	5F2	6C2
N23-89	6A1	6A2
R15-21	6A1	6B3-4
N20-66	6F5-7	7A6-8
R01-53	7A1	7B2-3
N26-01	7A6-7	7B2-3
N02-70	7B1	7B2
N12-93	7B1	7B4-5
N03-75	7C1	7C2
N09-15	7C1	7C3-4
N14-11	7D2	
N04-05	7D2	7D9
N09-26	7E1	7E2

YAC	START READ	END READ
E00-71	7E1	7E4
E02-82	7E1	7E6
N12-74	7F1-2	7F8-10
DY867	7F3-4	8B1-2
N10-55	7F7-9	8A5
R07-36	8A1	8C2-3
N04-21	8B3	8C1,2
E01-42	8C13	8D10,11
R16-01	8D10-11	8E6-9
R19-61	8E1-2	8F1-2
N03-36	8E3	8E12
N17-61	8E3-4	8F9-10
N23-10	9A1	9A3-4
R25-03	9A1	9A3-4
DY860	9A2	9A3
N13-46	9A3	9B1
DY886	9B1	9B14-15
R07-57	9B6-7	9D2-4
R15-64	9B8-10	9C1-2
R20-59	9C2-3	9E1
N26-20	9C3-5	9D3-4
DY806	9D1-2	9E1-2
Rt13-74	9D3-4	9F5-10
DY600	9E1	9E2
DY805	9E1	9F10-13
N03-66	9E4	10A1
E00-57	9F1	9F13
N04-80	9F1	9F8,9
R16-45	9F5-6	10A2-3
DY288	9F8-11	10B2-3
DY396	10A1	10A3
R18-90	10A2-3	10A10-11
N03-10	10A3	10A6,7
N19-75	10B7-8	10C2-4
N09-72	10B8-9	10C1-2
N24-12	10D2-3	11A1
RX01-96	11A6	11A10-11

YAC	START READ	END READ
N24-20	11A7-8	11B1-2
R21-50	11B17-19	11D1-2
N17-71	11D1	11D10-E1
N10-08	11D1	11D8-E1
R20-11	11F1-2	12A5-6
E02-22	11F3	12A1
DY618	12A1	12A2
DY766	12A2-3	12A9-10
R16-51A	12A8-10	12D2-3
R16-57	12A10	12D2
N06-21	12B1	12C7,8
R19-07	12B1	12C7-8
N12-25	12B2-4	12D1-2
DY106	12C1,2	12D2
R13-12	12C1-2	12D2-3
N04-06	12D1	12D2
E01-71	12E1	12E2
DY122	12E1	12E7
N19-21	12E5-8	12F2-3
R15-34	12E8-9	12F2-3
R16-37	12F3-4	13A10-12
R03-06	13A1	13B4-5
Rt05-24A	13A5-6	
E00-38	13B1,2	13C1,2
DY173	13C1,2	13D2
N03-21	13C1,2	
DY751	13D3-6	13E4-5
DY822	13E1-2	13F7-15
DY922	13E10-11	14A4-5
DY125	14A6,7	
DY699	14B1	14B15-18
DY802	14B1	14B3-4
R17-25	14B3-4	14C1-2
R16-08	14B5-6	14C6-7
N10-89	14F1-3	15B1
N03-17	15A6	15A11
N12-19	15D1	15F1-2

YAC	START READ	END READ
N12-21	15E7	16A1-2
N18-70	15F1-4	16A1-2
N16-96	15F2-4	15F5-6
N02-30	16A1,2	
DY451	16F1	16F7-17A1
R23-40	16F1	17A1-3
DY755	17A1-2	17A10-12
N12-44B	17A5-6	17B2-3
R25-10	17A7-8	17C2-3
N25-33	17A7-9	17B2-3
N03-34	17D1,2	17D3,4
N16-71	17D4-5	17F1-2
DY877	17D5-6	18A2-3
DY678	18A1	18A3-4
DY705	18A2-3	18B1-2
R03-28	18A5-7	18C4-5
N18-82	18B10-11	18C4-5
N11-04	18C1	18C4
N04-53	18F1	18F2
N15-63	18F4-5	19B2-3
R23-39	19A1	19A4-5
R07-48	19B1-2	19E5-6
R01-63	19B2-3	19E1-2
N14-15	19B3-C1	19C3-6
E02-33	19C1,2	19C6
N02-91	19C1,2	
R04-37	19D1-2	19E3-4
N27-47	19D1-2	19E4-5
N02-90	19E1	19E2
N18-52	19E1	19E4
N27-32	19E1	19E4
R21-76	19E2-3	19E7-8
DY582	19E5	19F2
N19-79	19F3-4	20A2-3
N05-87	20A1,2	
N16-73	20A1,2	
R19-85	20A3-5	20B-C
R18-12	20A5-B1	20C2-3
R14-43	21B2-3	21B4-7
N13-91	21B4-7	21C1-2
N01-70	21C1	21C3
N13-64	21C3-4	21D2
N19-86	21C7-8	21D2
DY818	21D1	21D2
E00-58	21D2	
DY213	21D4	21E1
N03-19	21E1	21E2
DY584	21E1-2	21F3-4
N03-64	21E2	
N10-22	21E2-3	22A1-2
DY541	22A1	22A3
Rt03-28	22A1-2	22A5-8
DY670	22A3	22A6
Rt04-22	22A5-7	22B6-7
R22-78	22A8-B1	22B7-8
N10-09	22B3	22B5
R02-46	22B8-C1	22E2-3
R11-45	22C1	22D4-6
DY202	22D1-2	22E3-4
N24-80	22E1	22F1-2
R19-63	22E1	22F2-3
DY642	22E1,2	22F2
N02-50	22E1,2	22F2
R16-92	22E2-3	22F3-4
R22-75	22F1	23A3
N09-42	22F1-2	23A1-2
DY730	22F1-2	23A5-7
DY612	22F3-4	23A2-3
R24-21	22F3-4	23B2-3
Rt02-64	22F3-4	23C2-3
R23-60	23A1	23A1-2
RX02-10B	23A2-3	23C1
N06-84	23C4	23E1,2
R21-72	23D1-2	23E3-4
DY469	23D3	23E4
N05-47	23E1,2	

YAC	START READ	END READ
E02-92	23F1,2	
R25-27	23F2	24A2-3
DY685	24B1	24C1-2
DY243	24B1,2	24B3
R11-16	24C1	24D2-3
DY801	24C1-2	24D1-2
DY617	24D1	24E1-2
R25-91	24D1-2	24E1
Rt06-05	24E1	25A3-4
N22-78	24E1-2	25A1
N17-23	24E2	25A2-3
DY10	24E5	24F6
N23-06	25A1	25A4
R17-30	25A1	25A5-8
R16-25	25A2-3	25B5-6
E02-38	25B1,2	25B10
R17-64	25C4-7	25D2-3
R25-59	25D1-2	25E2-3
N20-72	25E1	26A1-2
N23-08	25E1	26A8
N05-76	25F1	25F2
N25-90	25F1	25F4
N22-11	25F1	25F5-26A1
RX02-16	25F3-4	26A1-2
DY148	26A1	26A6
R19-62	26B7-8	26C3-4
R22-41	26C2-3	26D2-3
DY817	26D1	26E2
DY133	26D1,2	
R19-84	26D10-E1	26F5-7
N12-06	26D10-E1	27B1-2
R18-64	27A1	27C5-6
R18-09	27A1-2	27C1-2
E02-07	27C1	27C2
R24-23	27C8-9	27F1
N04-44	27E1	27E2
DY271	27E1	27E3
RX04-95	27E1	27E7-8
R16-84	27E3-4	27F2-3
R15-17	27F4-7	28B3-C1
R25-53	28A1	28B3-4
DY609	28C1-2	28D3-4
RX01-64	28D1	28D4-5
N03-69	28D8	28E4
R12-23	28E2-4	29A1-2
N13-72	28F3-5	29D1-2
N18-37	29D1	29E2
N13-38A	29D1	29E2-3
N02-31	30A3	30A6
DY893	30A3-4	30A8-9
R18-83	30A3-4	30A9-B2
DY761	30A6	30A8
R15-73	30A6-7	30B2-3
E03-40	30B5,6	30C5
DY918	30C1	30C5-6
N14-84	30C3-5	30F3-6
N14-85A	30C4	31A2
DY412	31A1	31C7
DY401	31A1	31D11
DY400	31B1	31C7
DY154	31C1,2	31D11
R12-90	32A2-3	32C1-2
N03-26	32B1,2	
E01-76	32C1,2	
N03-90	32E3	32F1
R14-08	32E3-4	33A2-3
E02-03	32F1	32F2
N06-11	32F1	33A2
R17-44	32F1	33A8-B1
N10-65	32F2-3	33A6-8
N05-59	33A1	33A2
N04-01	33A1	33A4
DY731	33A5-7	33B2-3
N12-68A	33A6	33C6
N25-47	33B1	33C1-2
N26-65	33D1-2	33D4-E2

YAC	START READ	END READ
DY919	33D1-2	33D3-4
R23-58	33D2-3	33E7-9
DY424	33E6	33E10
R04-79	33E6-8	33F2-3
E01-74	33E7	33E8
DY191	33F1	34A2
R24-35	33F1	34A5-7
DY220	34A1	34A2
DY691	34A1	34A2
R15-05	34A1	34A9-11
R12-28	34A1-2	34A2
N20-14	34A5-6	34B2-3
R11-31	34A5-6	34B4-5
R03-65	34A5-7	34B7-8
R05-80	34A6	34B2-3
N07-03	34A8	34B2
E03-30	34B6	34D1,2
Rt05-50	34B8-11	34C5-6
N05-63	34C1	34C4,5
N15-74B	34C3-4	34C5-6
N07-36	34C4,5	
E01-77	34E1	34E5
E02-09	34F1	34F4
R14-68	34F1-2	34F3-4
N05-34	35A1,2	
R15-19	35B2-3	35B6-7
N12-56	35B6	35B7
DY641	35B6-8	35B9-10
DY620	35C1	35C4
Rt03-48	35D1	35D4-5
R18-51	35D1	35D5-7
N09-47	35D1-2	35D3-4
DY784	35D5-7	35F1-2
DY681	35D6	35E3
DY668	35E1	35E4-F4
DY788	35E1	35F3
DY787	35E1	35F5
R14-44	35E1-2	35F2-3
N18-56	36A3	36A7-8
N06-08	36B1,2	
R14-83	36B8-C1	36D1
R07-11	36C1	36C2
R07-72	36C3	36D3
DY629	36D1	36D2
R07-63	36D1	36D2
R18-38	36D1	36D3-E1
R22-45	36D1	36E2-3
N11-41	36D3-E1	36E3-4
R07-37	36E1	36E5
N24-19	36E1	36E3-4
R19-59	36F1-2	37A2-3
R16-27	36F6-7	37B1-2
N02-89	37A1	37B1,2
RX02-37	37A1	37B4-5
N09-63	37A1-2	37B1-2
DY132	37B6	37C6
DY727	37C1,2	37D2
N19-85	37C5-7	37D5-E2
R17-14	37F6-38A1	38A7-8
DY67	38A1	38A2
N15-73	38A1	38A2-3
N13-28	38A1	38A5-8
N17-30	38A3-4	38B4-6
R15-50	38A4-5	38A8-B1
DY285	38B1	38B2
DY386	38C1	38D2
E03-06	38C1	38D2
DY399	38C5	38D2
DY385	38D1	38E3
R11-02	38D1	38E7-10
N21-29	38D1,2	38E4-7
E01-65	38E3	38E10
R18-74	38F2-5	39A1-2
DY250	39A3	39C4
N02-47	39B1,2	
DY917	39C1	39C2

YAC	START READ	END READ
R23-29	39E3-4	40A2-3
N12-87A	39E5-6	40A3-4
N03-35	40A1	40A4
N18-93	40A1	40A4
Rt02-72	40A1	40A4-5
N09-89A	40B1-2	40B3-4
E02-42	41C1	41F11
E02-85	41C1	41F11
E03-07	41C1	41F11
N10-43	41E3-5	41F1-2
N21-20	41E3-5	41F5-8
R16-15	41F1	41F2
N16-79	41F1	41F2-3
DY361	41F1	41F3
N10-17	41F1	41F3
N28-07	41F1-2	41F6-8
R19-21	41F2-3	41F10-11
RX02-33	42A1	42A2
E01-62	42A1	42A3
N03-62	42A1	42A3
N13-83	42A1-2	42A18-B3
R17-37	42A1-2	42A4-5
DY849	42A2	42A8-9
DY234	42A3	42A18,19
N15-67	42A4-5	42A8-10
DY398	42A6	42A12
N14-37	42A8-10	42A18-B1
R18-65	42A14-19	42B1-2
N16-37	42A15-17	42B3-4
DY798	42B1-2	42B4-C6
E01-57	42C1	42C10
N23-63	42C3	42E3
DY902	42C3-5	42D3
N06-18	42C6	42D4,5
R12-56	42C7-10	42D5-6
N03-92	42C8,9	42D6
N23-37	42C8-9	42E1
N11-82	42D4,5	42E2
N22-62	42E3-5	43A2-3
R16-60	42F1-2	43A2-3
DY169	43D1,2	43E9,10
N02-51	43D1,2	43E9,10
N06-06	43E1,2	43E6
N10-80	43E1-3	43E7-8
DY714	43E8-14	43F1-2
Rt01-41	43E14-17	44A2-3
E02-19	44A1	44B5,6
E02-10	44A4,5	44B4
DY619	44C1	44C2-3
DY281	44C1,2	44D2
N10-10	44C1-2	44D1-2
N09-45	44C3	44D3
N05-23	44C4,5	44D2
E01-80	44C4,5	
DY898	44D1	44D3-4
N05-96	44D4	44F1
N22-41	44E1-2	44F3-4
N12-63	44E1-3	44F1-2
DY12	44F1	44F2
N26-90	44F2-3	45A2-3
DY242	44F9,10	45A3
N24-41	45A1	
R19-72	45A2	45C1-2
N20-67	45C1	45E1-2
N19-88	45C3	45D10?
N14-48	45C7-9	46A3-4
N02-21	45D1,2	45E1,2
R25-46	45D1-2	45F1-2
N14-49	45D5	46B1,2
N24-84	46A1	46A3-4
N17-62	46A1	46A4
DY114	46B1,2	
N02-52	46B1,2	
DY626	46B3-4	46D2-3
N03-73	46B10,11	46D7
N16-40	46C11-D1	46F6-7

YAC	START READ	END READ
R15-68	46E7-9	47A2
DY937	46F7	46F10-11
R14-41	46F10-47A1	47A5-8
DY19	47A11,12	47B4,5
R25-11	47A13,14	47D6
R03-20	47B4	47B7-8
DY728	47B4	47D1
N03-38	47C1,2	47D4
R17-69	47D1-2	47D5-6
R19-34	47F6-F1	48A1
R24-76B	47F3-5	48A2-3
N02-76	47F4	47F18
R17-63	47F10-14	48A4
N03-81	48A3,4	48C1
N14-62	48A3-4	48B6-7
R17-49	48A5-B1	48C5-6
R19-41	48B1-2	48C2-3
R03-43	48C1	48C4-7
N10-34	48C1	48C5-8
N15-20	48D1	48E2-3
N12-62	48D2-3	48E1
N17-64	48D3	48E2
R23-77	48D6-7	48F2-3
R19-54B	48D8	49A6?
DY124	48E1	48E2
DY179	48E1	48E3
R19-66	48F2-3	48F6-7
DY184	49A1	49A2
R16-41	49A3-4	49B10-12
DY878	49A9-11	49B1-2
R06-67	49C1-2	49E1-2
N25-60	49D1	49E7
N12-92A	49D1	49E7-F1
N25-60	49D1-2	49E6-7
R05-85	49E5-7	49F4-6
E01-44	49E6	49F2
R12-14	49F3-4	50A3-4
N05-43	49F7,8	

YAC	START READ	END READ
DY343	49F10	50A3,4
DY231	49F13,14	50A1,2
N03-45	50A1,2	50A3,4
N10-95	50A5-7	50B6-9
N04-87	50A9,10	
R16-07	50A10-11	50C3-4
N09-10	50A12-14	50C4-5
N17-88	50B1	50C1-2
N27-48	50B1-2	50C3-4
N21-05	50C1	50C10-12
DY287	50C1	50C3,4
N18-25	50C1	50C4-5
N17-22	50C8-9	50C23-D1
N14-27	50C9-11	50D2-3
Rt01-57	50C14,15	50E8
N17-51	50D1	50E6-7
R07-34	50F1-4	51A5-6
N03-28	51A1	51A2
N09-90	51A6-7	51C1
N09-34	51A7	51B9
N28-28	51B1	51C2-3
N06-37	51D6	51D7
N20-38	51D6	51E7
N20-27	51D9-11	51E2-3
R15-51	51D9-11	51E6-7
N17-08	51E1	51E5-6
R15-63	51E1	51E7
N04-84	51E5,6	51F1,2
R07-31	51F1-2	52A5-7
N22-16	51F7-10	52B3-4
N16-65A	52A4	52B2-3
DY18	52B1	52C2
N28-02	52B1-2	52C6-9
E01-67	52C1	52C5
DY551	52C1	52D1-2
R07-60	52C1-2	52D1-2
N27-23	52C1-4	52C8-D2
DY517	52C8	52E9

YAC	START READ	END READ
DY522	52C9	52D15
R17-20	52D1-2	52E1-2
R15-76	52D7-8	52E9-11
DY101	52D9,10	
DY773	52E1,2	53B4
N20-75	52E5,6	52F4
N03-74	52E7	52F8
R22-59	53B3-C1	53C7-11
N07-14	53C1	53C4
R24-56	53C1	53C8-11
N21-47	53C1	53D3
R21-11	53C1-2	53D2-3
DY460	53D1	53E4
R18-80	53D10-11	53E2-3
R14-17	53D10-13	53F1-2
N13-66	53D12-14	53E9-11
DY11	53E4	53E11
DY420	53F1	53F2
N12-75	54A1	54B2-3
N28-01	54A2-B1	54B16-18
DY283	54B1	54B10
R06-45	54B1-2	55D1-2
R24-60	54B10-16	54D2-3
DY417	54E4	55B2,3
DY418	54F1,2	55A2
DY282	55A1	55A3,4
DY843	55A1	55A4-B3
DY139	55A1-2	56B1-2
R06-51	55B3-4	55C6-7
N22-40	55B5-10	55C5-8
R17-59	55C1	55C5
N05-30	55C1	55E12
R03-14	55C6-8	55E2-3
DY57	55C11,12	55F1
N19-69	55F1-2	56B1-2
N03-51	56A1	56A3
E01-86	56A1	56B2
DY848	56A2-3	56B1-2

YAC	START READ	END READ
R22-48	56B5,6	56C9
E02-26	56C9	56D2
N10-14	56D1	56E3-5
N22-88	56D10	56F12
N25-59	56D11-E2	56E5-6
N04-47	56E1	56E2
R18-15	56F2-3	56F7-8
N26-70	56F4-6	56F13-15
R21-33	56F8-9	56F16-17
RX01-81	57A1	57A4
R17-43B	57A5-6	57B3-4
DY586	57B1-2	57B5-6
R18-77	57B4-5	57B10-13
N17-58	57B4-5	57C1-2
N03-49	57B7	57C2
R24-34	57B8-11	57D2-3
N03-18	57B13	57B14
N04-76	57B13	57C1
R23-92	57C2-4	57E2-4
N05-09	57D11	57E6,7
R15-93	57D11-E1	57F2-3
N28-96	57E1	57E2
R17-75	57E1	57F4-7
N14-34	57E5-7	57F6-8
N24-26	57F8-9	58A3-4
DY277	58A1	58A2
E01-81	58A1	58A2
N24-15	58A1	58A4
RX02-27	58A2-3	58B6-7
DY741	58B3-4	58D2-3
DY739	58B6-7	58D2-3
R12-33	58C1-2	58D2-4
Rt02-43	58D1	58D7-8
R23-17	58D3-4	58D7-8
N19-77	58D6-E2	58F4-7
N11-66	59A1	59A3-4
R22-60	59B4-6	59C3-5
DY410	59C1	59C4

YAC	START READ	END READ
N26-58	59C1	59C4
R25-101	59C4-5	59D3-4
N18-87	59D7-10	59F2-3
E02-30	59F1	59F3
E02-16	59F5	60A2
E02-01	60A1	60A2
R25-80	60A8-11	60B10-12
N13-25	60A10-13	60B12-13
N11-96	60A15-16	60B11-12
N04-02	60B1	60C1
E01-49	60C1	60C2
N04-83	60C7	60D2
N01-71	60D7	60D9
N21-10	60D9-10	60D14-15
N28-18	60D14-16	60E7-9
N02-74	60E1	60E4
DY226	60F1	60F5
DY68	60F1	60F5
N06-24	60F1	60F5
R11-32	60F1-2	60F3
DY909	60F2	60F3
N23-66	61A1	61A5-6
N02-24	61C3	61D4
N28-17B	61C4-7	61D1-2
R07-14	61D1	61D2
N03-94	61E1	61F1
N11-70	61E1	61F1
N10-01	61E1	61F2-3
N16-88	61E1	61F2-3
N14-05	61F1-2	61F4-5
N13-38B	62A1	62A11-12
DY149	62A1	62A3
N14-43A	62A1-2	62B3-6
N28-67	62B1-2	62B10-11
N18-02	62B2-4	62C2-3
R18-25	62B3-4	62B11-12
N23-77	62B10-11	62C2-3
DY134	62D1	62D2
DY402	62D5	62E2
DY903	62D5	62E2-3
DY759	62D5	62E3
R07-10	62D5-6	62E2-3
RX04-38	62D5-6	62E4-7
R11-20	62E5-7	62F2-3
N13-19	62F1	63A1-2
DY753	62F4-5	63A1
N21-67	62F4-5	63B2-3
DY406	63A1	63A3
DY177	63B7	63C1
R16-17	63C1	63D2-3
N13-93	63D2-3	63E5-6
R18-53	63E1	63E3-4
R21-49	63E1	63E3-4
R18-72	63E1	63E5
DY527	63E1	63E5-6
R11-21	63E2-3	63E5-6
DY839	63E4-5	63F6-7
R22-28	63E4-5	64A2-3
DY750	63E6	63F3
N09-76	63F1	64A4-5
DY895	64A3-4	64A10-B1
Rt03-73	64A7-9	64B10-11
R04-91	64B1	64B9-10
DY276	64B9,10	64B17
N22-31	64B12	64B13
N25-23	64B12	64B16-17
DY669	64C1	64C4
N06-59	64C1	64C4
R15-89	64C4	64C5
E01-16	64C8	64C15
DY824	64D1	64D2
E01-32	64D1	64D2
DY604	64E2	64E12-13
N07-22	64E2	
R13-33	64F3-5	65A2-3
R25-40	65A7-8	65B3-4

YAC	START READ	END READ
R13-64	65B1-2	65C3-5
N05-89	65B4	65C1,2
Rt09-17	65C3-4	65E2-3
N20-69	65D1	65D3-4
DY405	65D1	65E2-3
DY241	65E1	65E10
N03-61	65E1	65E4
DY845	65E10-11	66A1
N13-52	65E10-11	66A2-3
R24-41	65E11-12	66A1-2
N09-59	65F1	66A3
N21-49	66A1	66A10
R15-75	66A1	66A5
N28-81	66A1	66A5-6
N25-66	66A1	66A6-9
N03-95	66A2	66A12
R01-57	66A5	66A8-9
DY278	66A10	66A18
R11-19	66A16-18	66B2-3
DY147	66A17	66B7-9
DY135	66B1	66B13
DY549	66B1	66C2-3
R15-22	66B1-2	66C2-3
N11-47	66C4-5	66D2-3
DY712	66C8	66D4-6
N14-76	66C8,9	66D1,2
N12-30	66C11	66D1,2
N02-63	66D1,2	66D10
N12-20	66D13-15	66E2-3
N23-90	66D14-15	66E2-3
DY178	66D15	66E2
DY2	66E1	66E6
DY697	66F1	66F5-6
DY259	66F1	67A1,2
DY232	66F4	67A4
DY543	67A1	67B1-2
N12-72	67A1	67B5-8
N09-40	67A7-9	67B7-9
R21-93A	67B9	67C5
R20-44	67B9-11	67C3-4
N03-83	67B10	67C4
N02-48	67C1	67C4
N04-67	67C1	67C4
Rt05-60	67C7-9	67D7-8
DY121	67D1	67D2
R04-93	67D3-4	67D11-13
N04-07	67D8	
N28-09B	67D8-9	67D12-13
E02-04	67D9	67D10
E02-06	67D9	67D13
DY190	67D11	67E2
DY854	67D13-E1	67E4-5
E01-87	67E1	67E3
DY150	67E1	67E4
R21-35	67E2-3	67F4
N27-12	67E3	67E4
N11-06	67E6-7	67F3-4
E01-84	67F1,2	67F4
R25-35	67F2-3	68A2
N25-53	68A1	68A2
R12-60	68A1	68A3-4
N03-82	68B1,2	68C4
DY661	68C5	68C12
DY904	68C7-8	68D3-4
DY743	68D4	68E3
R12-07	68D5-6	68E3-4
DY293	69A1	69C1
DY716	69A2-3	69B1-2
N12-70	69B1	69C7-10
N15-52	69D2-3	69F4-6
N27-17	69D3-4	69F2-3
DY235	69E6,7	69F2
R15-23	69E8-F1	69F7-70A1
N06-09	69F1	69F2
DY833	69F2	70A7
N17-42B	69F2-3	69F4-7

YAC	START READ	END READ
DY726	69F2-3	70A3-7
R19-03	69F3-5	70A1-2
R17-38	70A1	70A4-5
R12-01	70B1-2	70C1
N20-11	70C1	70C11-13
R23-14	70C1	70C2
N16-13	70C2	70C15-D1
R19-51	70C2	70C3
N16-14	70C2	70D3
DY673	70D1	70D2
N06-35	70D4	70D5
N03-71	70E4	70E8?
E01-53	70F1,2	
N04-63	70F1,2	
DY267	70F5-6	71C1
R14-37	70F5-7	71A3-4
N04-68	71A1	71A2
R25-17	71A1	71B1
R15-65	71A2	71B1
N11-87	71A3-4	71B1-2
R21-66	71A4	71C1
N06-90	71B1	71B2,+
R24-54	71B2	71C1-2
R07-23	71B2	71C2-3
N02-22	71B2,+	71B4
N24-27	71B4-8	71C1-2
E01-04	71C1	71C2
N12-44A	71C1	71C3-4
N20-96	71E1-3	71F2-3
R18-89	71E3-5	71F4-5
E02-35	71E5	71F2
DY894	72A1	72A2
N05-19	72A1	72C1
R19-02	72A2	72D2-10
R21-03	72C1	72E2-3
N12-07	72D1,2	72E4
N10-66	72D4-8	72E2-3
N27-06	72D5-7	72E1-2

YAC	START READ	END READ
N19-57	72D6	73A7
R23-13	72D8-10	72E4-5
N03-12	72D10,11	72D12
DY476	72D10,11	73A4,+
R14-26	72E1	72E3-4
R17-06	72F3-4	73C1-2
DY163	73A1	73A4,+
R15-53	73B3-4	73C2-3
R15-57	73C2-3	73D5-7
N13-77	73D2-3	73E4-6
DY847	73E3-4	74A2-3
R24-32	73E5-F2	74A2-3
Rt01-02	73F3-4	74A6-B3
N26-11	74A1	74A5
R19-43	74A2-4	74D1-2
R25-94	74C1	74F1
N14-88	74D1	74E3-5
R18-01	74D1	75A1-2
DY224	74D1	75F2
R11-28	74D3-5	75A4-7
R21-18	75A1	75A9
RX04-81	75A1-2	75B4-7
N15-29	75A7-8	75B8-10
N09-21A	75B1	75B5-7
R12-85	75B5-7	75C1-2
DY187	75B6	75C6
E01-70	75B10	75C2
R17-53	75B10-12	75C2-3
DY657	75C1	75C3-5
DY746	75C1	75C6
R24-53	75C2	75C5-7
N03-96	75C6	75D2
N15-10A	75D3-5	75F3-5
DY3	75E1	75F1
N19-60	75E1-2	75F7-9
N03-01	75E2	75F8
R22-39	75F3-4	76A4-5
R17-60	75F3-5	76A5-6

YAC	START READ	END READ
N01-15	75F6	76A1
N26-35	76A1-2	76B2-3
N05-37	76A3,4	76A6
R19-39	76B1-2	76B8-10
N12-69	76B3,4	
N07-56	76B5	76B11
N09-46	76B8-11	76D2-4
N06-20	76C1	76D3
N09-31	76C1-2	76D5-8
N03-40	76D6	76F3
R24-57	76E1	76F2-3
R20-43	76F1	77A2-3
DY485	76F1	77A3-4
Rt02-69	76F3-77A2	77B1-2
R04-13	77A1	77B2
DY317	77A1-2	77C5-7
E01-41	77C1	
N03-91	77D1,2	77E4
Rt04-04	77D3-4	77F3-5
N20-51	77E1	77F2-5
DY653	77E1-2	77F4-5
DY837	77F3-4	78A1
N06-54	78A+,1	78A2,+
N19-03	78A+,1	78A3
N18-35	78A1	78A3
R15-38B	78A1	78B3-4
R15-37	78A2-3	78B3-4
N24-46	78B1	78C1-2
N18-28	78B1	78C2-3
N12-02	78C1	78D3
E00-64	78D5	78F1
DY175	79A1	79A4
N18-84	79A1	79C1
N17-72	79A2-3	79C1-2
E02-41	79A4	79A5
N18-50	79B1	79B3
E00-62	79C1	79C2
R14-49	79C1	79D1
N13-11	79C2	
Rt15-11	79C2-3	79E1-2
N12-57	79D1	79E4
R16-55	79D2-3	79E4
R16-54	79E1	79E2
N13-65	79E1	79E4
DY442	79E1	79E6-8
R04-69B	79E1	79F1-2
DY7	79E3	79E4
DY429	79E3	79E8
N17-13	80A2	80C3
DY797	80A3	80B3
N02-72	80A4-B1	80B3
N10-90	80B1	80B2
N21-82	80B1	80B2
DY98	80B1	80B3
DY246	82A1,2	82A5
N02-33	82A1,2	82B4
N11-39	82C1	82C2
N01-68	82C1	82C3
DY251	82D1	82E8
N10-45	82D5-7	82E4-6
N16-86	82D7-8	82E3-5
N15-42	82E1	82E3-4
N16-72	82E3-4	82E7-8
N12-31B	82E4-5	82E7-8
DY404	82E6	82E7
N20-40	82E6-7	82F2-3
DY710	82F1	82F8-9
N24-50	82F1	82F8-9
N02-25	82F9	83A6
DY207	83A1	83A8-B2
E03-32	83A1	83A9?
R24-73	83A1	83B4-8
N03-37	83A5	83C6
N09-14A	83C4-7	83E2-3
N06-49	83C6	83D2
N15-43	83C7-8	83D2-3

YAC	START READ	END READ
N09-14B	83C8-9	83D5-E1
E00-63	83D1	83D2
R17-05	83D1	83E6
R21-59	83D2-4	83E2-3
N03-08	83D3	83D5
N04-28	83D4	83D5
N12-12	83D4	83D5
Rt01-08	83D4-5	83E3-5
DY622	83E2-3	84A3-4
N06-67	83E4	84A1
N28-05	83E4-7	84A2-3
DY482	83E6	84A2
N20-63	83F3-4	84A5-B1
R13-09	84A1	84A2-3
N28-59	84A1	84A5-6
E03-42	84A5,+	84B2
N10-64	84A5-6	84B2-3
R19-87	84B1	84C3-4
N03-05	84B1	84C6
R06-93	84B1-2	84D1-2
DY896	84B2-3	84B4-6
R19-94	84B2-3	84D1-2
R22-23	84C3-4	84D11-12
DY624	84D1	84D2
N05-64	84D1	84D2
N19-33	84D1	84D5-8
DY219	84D3,4	84D10
R14-02	84D8-9	84E2-3
N03-42	84D11	84E1,2
N28-44	84D11-12	84E4-6
N28-38	84E8	84F2
N23-73	84F1	84F2
N24-53	84F1	84F2
N18-68	84F2-3	85A3-4
N07-17	84F6,7	85A1
DY557	84F8	85A1
DY841	85A1	85B1-2
N19-19	85A1	85B5-6

YAC	START READ	END READ
DY863	85A2-3	85A10-11
R20-65	85B1-2	85C2-3
N02-20	85B8	85B9
N12-60	85C3-4	85C11-D1
N03-43	85C6	85C13
R18-68	85C8-11	85D2-3
N19-46	85D1	85D2
DY115	85D1	85D4,5
DY390	85D1	85D4,5
DY914	85D1	85D6
DY388	85D1	85D7-10
R14-96	85D1	85D9-10
E00-35	85D4,5	85D6
R19-73	85D11-12	85E2-3
N06-02	85E4	85E12
N04-25	85E4	85E5
DY782	85E12-14	85F2-3
R17-61	86B3	86C1
Rt01-39	86B4-5	86C8-11
R23-07	86B6	86C5
Rt03-30	86C1	86C7,8
R22-61	86C2	86C5
N05-93	86C4	86D1
R18-81	86D1	86D2
E02-25	86D1	86D5
N27-13	86E1	86E2-3
N13-92	86E3-4	86E11-15
N27-45	86E7	86E10-12
R24-11	86E15-18	87A1-2
N27-05	86F2	87A5-6
R21-15	87A8-9	87B3-4
N11-43	87A9	87B9
DY136	87C1-2	87D1
E01-28	87C4	87C8
DY313	87C4	87D6
N25-54	87C6-7	87D4-5
DY280	87D1	87D4
N02-49	87D1	87D5

YAC	START READ	END READ
DY814	87D1-2	87D11-14
DY183	87D3	87D4
N06-34	87E1	87E2
DY475	87E1	87F2
R23-78	87E4-5	87F2-3
N12-76	87F1	87F3
R18-36	87F3-4	87F14-15
N15-82	87F11-12	87F14-15
N21-08	87F11-12	88A1
DY239	87F12	87F15
DY238	87F12	88A4,5
DY930	87F12	88A7
DY707	87F12	88A8
DY913	87F12-13	88A1
R22-54	88A9	88B7
R23-12	88A10	88B6
DY857	88B1	88B6-8
R25-06	88B7-9	88C9-10
DY236	88C1	88C6
DY568	88C3-4	88C9-D2
R19-52	88C6-7	88D5-6
N24-29	88D8	88E2-3
E03-09	88D8	88E4
N20-94	88D10-E1	88E3-4
N15-89	88E4-5	88F1-2
R12-08	88F5-8	89A1-2
N11-16	88F8-9	89A1-2
N15-10B	89A1	89A3-4
N10-42	89A2-3	89A10-13
R14-86	89A8-11	89B7-8
N04-79	89B1	89B3
R15-24A	89B1-2	89B9-11
R20-47	89B2-3	89B16-18
DY230	89B12	89B20
N28-23	89B12	89C3
N26-85	89B14-16	89C2-3
DY713	89B17-19	89C6-7
R23-66	89C1	89D3-4

YAC	START READ	END READ
N05-05	89D1	89D9
N25-72	89D1	89E3-4
N10-59A	89D1-2	89D3-4
N10-13	89D3-4	89E1-2
R19-19	89D3-4	89E1-2
R21-81	89D8-9	89E3-4
N02-14	89E1	89E4
N06-28	90A1	90A2
R23-64	90B1	90C4-6
R18-79	90B1	90C7
N16-12	90B1,2	
N13-62	90C1-2	90E2-3
N03-31	90C4	90C7,8
N09-86	90D1-2	90E2-3
R22-17	90D2-3	90E2-3
R06-17	90D3-5	90E3-5
DY702	90E1	90F1-2
N14-14	90F1-2	91A2-3
N26-17	90F9-11	91A2-3
N24-21B	91B2-3	91B5-6
DY261	91B6	91C4
DY565	91C4-5	91E4-5
N03-88	91D1	91D2
N09-75	91D1-2	91E1-2
DY891	91D2	91F1
R21-05	91D2-3	91F1-2
R18-05	91D3-5	91F2-3
R11-07	91F1	91F11-12
DY166	91F1	91F4,5
N13-74	91F1-2	92A1
E01-01	91F4,5	91F11
R16-29	91F4-5	92A4-5
R11-05	92A1	92A7-8
DY605	92A3	92A6
N07-75	92A6	92C4
N09-29	92A9-11	92B3-4
N03-79	92A10	92B1
N04-14	92B1	92B3

YAC	START READ	END READ
R16-23	92C2-3	92D7-9
DY876	92C3-5	92E1
N21-58	92D1	92D2
N11-84	92D1	
N18-32	92D2-3	92E2-3
N19-53	92E1	92E10
N05-11	92E1	92E2
N25-57	92E1	92E8-9
R07-35	92F1	92F5-7
N19-48	92F1	92F8-10
DY44	92F1	93A1
R14-74	92F1-2	93A2-3
R14-11	92F5-6	93A4-5
R23-50	93A2-3	93B9-12
DY126	93B4	93D9
N26-73	93B6-C1	93D1-7
R11-33	93E1	93F1
R12-43	93E1-2	93F1
R18-47	93E2-3	93E9-F1
N09-35	93E4-5	93F4-5
DY667	93F1	93F4,5
RX04-08	93F8-9	94A3-4
DY655	93F9-10	94A3-4
DY792	94A1	94A3
DY411	94A1	94A4
N24-49	94A1	94A7-10
DY421	94B1	94B6
N04-57	94B1	94B7
Rt01-44	94B8-10	94D4-5
DY643	94C1	94D2-3
N19-81	94D1	94D10-E5
N17-28B	94D6-8	94D4-5
R22-79	94D8-9	94F1
E03-10	94F1,2	95A2
E03-19	94F1,2	95A2
DY286	95A1	95A10
DY54	95A1	95B5,6
DY107	95A4,5	95B1,2
DY852	95C3-4	95E1-2
N27-18	95C10-13	95D8-10
N26-74	95E1	95F1-2
R17-87	95E5	95F8-10
Rt02-35	95F1-2	96A2-3
N05-58	96A2	96A8
E01-79	96A2	
Rt02-47	96A3-5	96A20-21
Rt12-13	96A8-9	96B5-7
N21-80B	96A13-15	96A19-20
N03-44	96A17	96A20
N13-63	96A22-24	96B20-C1
R25-45	96B9-10	96C1-2
N03-56	96B10	96B11
DY804	96B14-18	96C7-9
N04-04	96B19	96C9
DY105	96C1	96C7
N13-54A	96C8-9	96D4-6
N28-08	96C8-9	96E1
DY715	96D1	96D2
E03-08	96E8	96F10,11
DY227	96F1	96F5
DY408	96F1	96F5
R21-90	96F7-8	97A6-7
R22-90	96F7-8	97A7
DY118	97B1	97B2
R24-65	97B1	97B8-9
N05-20	97B2	97C1
N02-17	97B4	97C5
N13-03	97B6	97C2-3
R25-38	97B6	97C4-5
R23-33	97C1	97D1
N04-56	97C1	97D2
R24-44	97E1	97F1-2
N18-11	98A1	98A14-15
N16-60	98A1	98A4
N18-08	98A1	98A4
RX01-49	98A5-8	98B2-3

YAC	START READ	END READ
R02-69	98A10-11	98B2-3
R25-42	98A13-15	98B8-C1
DY440	98B1	98B2
DY899	98B1	98B2-3
E01-08	98B1	98B4?
N15-15	98B1	98B8-C1
N23-56	98B1	98C1-2
N04-18	98C1	98D2
DY528	98C1	98D2-3
DY176	98C2	98D1
R13-80	98C5-D1	98D5-7
N04-03	98D1	98D2
DY123	98D1-2	98E3-6
DY5	98D2	98E1
N01-24	98D3	98E2
N17-96	98E1	98F6,7
N22-18	98E1-2	98F4-5
DY116	98F1	98F2
DY851	99A1	99A8-9
N05-51	99A1,2	99A5,6
N18-17	99A1,2	
R15-41	99A3-4	99A10-11
R15-12	99A4-5	99A7-8
N22-75	99B1	99B5
R11-22	99B3-4	99C1-2
Rt01-01	99B4-5	99C1
E01-56	99B7	99C2
R16-22	99B9-10	99D3-4
N05-72	99C1	99D1,2
N04-74	99D1,2	

YAC	START READ	END READ
R21-61	99D1-2	99E3-4
N03-48	99E1	99F2
DY162	99F1	99F11
N13-50	99F8-10	100A2-3
R25-37	99F9-11	100B1
DY704	100A1	100A5-6
DY815	100A1	100A6
N04-88	100A3	100A7
R15-71	100A7-B1	100B4
R13-77	100A7-B1	100B4-5
DY21	100B1	100B5
N25-44	100B4	100B5
N03-39	100B4	100B5
R07-38A	100B4-5	100C2-3
N17-87	100B4-5	100C4-5
R25-85	100B5	100C3
R25-84	100B7-9	100C2-3
DY419	100C1,2	100C4,5
Rt03-80	100C3-4	100D3-4
R16-72	100C4-5	100D3-4
R21-17	100C7	100E1
Rt02-03	100D1	100E2-3
N19-58	100D3-4	100E2-3
R16-53	100D3-4	100E2-F2
R14-90	100E2-3	100E3-F2
RX04-85	100F1	100F2-3
E01-75	102A1,2	102A8
DY35	102B1,2	102B7
N03-89	102F8	

APPENDIX 3

P1 LOCALIZATIONS

The tabulation here includes the estimated extent of euchromatic bands covered by P1 clones of the types described in chapter 2. Clone designations are given according to plate-well position in the Cambridge library ("CAMBR"), according to plate-well position in the Berkeley library ("BERK"), and according to the DS number ("DS No."). Plates 1 through 40 (in the Cambridge nomenclature) are in the vector pNS582-tet14 Ad10; the rest are in the vector pAd10 sacBII. These correspond, respectively, to clone numbers 1 through 40 and 41 through 96 in each plate in the Berkeley library.

Throughout we have used the band designations of Heino TI, Saura AO, Sorsa V. Maps of the salivary gland chromosomes of *Drosophila melanogaster. Drosophila* Information Service 1994; 73:619-738. Comments such as "band designation from Sorsa" refer to this publication. A number of the band designations include a question mark, which does not mean uncertainty in the cytology but rather uncertainty whether the band is the same as that described by Bridges.

In the annotations, terms like "secondary sites" and "weak sites" refer to hybridization signals most of which probably arise from transposable elements present in multiple copies in the genome. The term "CC" stands for chromocenter. A number of clones were also identified by library screening in preliminary work to determine the completeness of the library and the efficiency of contig assembly with STS mapping. The screening was carried out with anonymous cDNA clones from the adult head (specifically from head-not-embryo cDNAs) in a library described in Palazzolo MJ, Hyde DR, VijayRaghavan K, Mecklenberg K,

Benzer S, Meyerowitz E. Use of a new strategy to isolate and characterize 436 *Drosophila* cDNA clones corresponding to RNAs detected in adult heads but not early embryos. Neuron 1989, 3:527-539.

Most clones for which both the Cambridge and the Berkeley designations are given were examined by in situ hybridization using an isolate from the Cambridge library prior to the time that the Cambridge library was restreaked and reorganized into the Berkeley library. Clones for which the Cambridge designation is blank were examined by in situ hybridization using an isolate obtained directly from the restreaked Berkeley library. Because an occasional well in the Cambridge library may contain two different clones, a clone from the Berkeley library that is ostensibly identical to that in the Cambridge library may, in fact, be different. In such unlikely cases, the original clone with the desired location may still be recoverable from the Cambridge library. As noted in chapter 5, a total of 214 hybridizations were carried out in duplicate as blind controls on the repeatability of the localizations. Most of these were from Berkeley plates numbered 1 through 9.

For further information, see the references below or establish contact electronically with ftp.bio.indiana.edu in the directory flybase or flybase@morgan.harvard.edu. P1 clones are available on request from any of the following sources. When requesting clones, please provide both the plate-well location in the Berkeley library (for ease of recovery) and the DS number.

Spyros Artavanis-Tsakonas, Yale University, Howard Hughes Medical Institute, Boyer Center for Molecular Medicine, 295 Congress Avenue, New Haven, CT 06536-0812
Tel: (203) 737-4466, Fax: (203) 787-3364

Michael Ashburner, Department of Genetics, Cambridge University, Downing Street, Cambridge CB2 3EH, England
Tel: 44 223 333 969, Fax: 44 223 333 992

Hugo Bellen, Baylor College of Medicine, Howard Hughes Medical Institute, One Baylor Plaza, Room T634, Houston, TX 77030
Tel: (713) 798-5272, Fax: (713) 797-6718

Sean Carroll, University of Wisconsin-Madison, Howard Hughes Medical Institute, Laboratory of Molecular Biology, 1525 Linden Drive, Madison, WI 53706
Tel: (608) 262-3203, Fax: (608) 262-4570

Daniel L. Hartl, Harvard University, Department of Organismic & Evolutionary Biology, 16 Divinity Avenue, Cambridge, MA 02138
Tel: (617) 496-3917, Fax: (617) 496-5854

Steven Henikoff, Fred Hutchinson Cancer Research Center, Howard Hughes Medical Institute, Department of Genetics, Room A1-111, 1100 Fairview Avenue North, Seattle, WA 98109
Tel: (206) 667-4514, Fax: (206) 667-5889

Gary H. Karpen, Salk Institute, Department of MBVL
PO Box 85800, San Diego, CA 92186-5800
Tel: (619) 453-4100, X473, Fax: (619) 457-4765

Thomas C. Kaufman, Indiana University, Howard Hughes Medical Institute, Department of Biology, Jordan Hall, Room A-507, Third Street and Faculty, Bloomington, IN 47405
Tel: (812) 855-3033, Fax: (812) 855-2577

Paul Lasko, McGill University, Department of Biology, 1205 Ave Docteru Penfield, Montreal, PQ H3A 1B1, Canada
Tel: (514) 398-6721, Fax: (514) 398-5069

Marek Mlodzik, European Molecular Biology Laboratory, Differentiation Programme, Meyerhofstr. 1, D-6900 Heidelberg, Germany
Tel: 49 62 21 387 303, Fax: 49 62 21 387 306

Gerald M. Rubin, University of California at Berkeley, Howard Hughes Medical Institute, Department of Molecular & Cell Biology, Room 539 LSA Bldg., Berkeley, CA 94720
Tel: (510) 643-9945, Fax: (510) 643-9947

Matthew Scott, Stanford University School of Medicine, Department of Developmental Biology, Beckman Center, B300, Stanford, CA 94305-5427
Tel: (415) 725-7680, Fax: (415) 723-9878

Allan Spradling, Carnegie Institute Washington, Howard Hughes Medical Institute, Department of Embryology, 115 West University Parkway, Baltimore, MD 21210
Tel: (410) 554-1221, Fax: (410) 243-6311

Carl Thummel, University of Utah, Howard Hughes Medical Institute, Department of Human Genetics, Eccles Institute of Human Genetics Bldg. 533, Room 2100, Salt Lake City, UT 84132
Tel: (801) 581-2612, Fax: (801) 581-5374

Tsuneyuki Yamazaki, Department of Biology, Faculty of Science, Kyushu University, Fukuoka 812, Japan
Tel: 092-641-1101, Fax: 092-632-2741

S. Larry Zipursky, University of California at Los Angeles, Howard Hughes Medical Institute, 5-748 MacDonald Bldg., 10833 Le Conte Avenue, Los Angeles, CA 90024-1662
Tel: (310) 825-2834, Fax: (310) 206-3800

REFERENCES

Hartl DL, Nurminsky DI, Jones RW, Lozovskaya ER. Genome structure and evolution in *Drosophila:* Applications of the framework P1 map. Proc Natl Acad Sci USA 1994; 91:6824-6829.

Smoller DA, Petrov D, Hartl DL. Characterization of bacteriophage P1 library containing inserts of *Drosophila* DNA of 75-100 kilobase pairs. Chromosoma 1991; 100:487-494.

P1 LOCALIZATIONS

CAMBR	BERK	DS No.	BEGIN	END	COMMENTS
	8-27	DS00699	1B1	1B2	
	9-13	DS00781	1B2	1B3	MAYBE TO 1B1 &/OR 1B4 + WEAK AT 10E4;10F1
64-41	41-64	DS03904	1B5	1B9	MAYBE TO 1B10
	9-88	DS00856	1C4	1C5	ALSO SOME SECONDARY SITES
	9-4	DS00772	1D1	1D2	ALSO ~100 SECONDARY SITES
	9-11	DS00779	1D1	1D2	
48-16	16-48	DS01488	1D1	1D2	ALSO CC LIGHT + ~150 EUCHROMATIC SITES
135-26	26-92	DS02492	1D1	1D2	
65-46	46-65	DS04385	1D1	1D2	SAME AS DS08417?
65-88	88-65	DS08417	1D2	1E1,2	MAYBE TO 01D1; SAME AS DS04385?
	7-11	DS00587	1D3	1E1,2	
136-19	19-93	DS01821	1E1,2	1E3	MAYBE TO 1E4 + CC MODERATE + ~100 SCNDRY SITES
81-67	Not present	DS09283	1E1,2	1E3	MAYBE TO 1E4
62-61	61-62	DS05822	1E1,2		
	8-80	DS00752	1F1	1F2	ALSO CC MODERATE + SOME SECONDARY SITES
137-12	12-94	DS01150	2A1	2A2	
45-63	63-45	DS05997	2A1	2A2	
137-87	87-94	DS08350	2A1	2A2	
81-82	Not present	DS09298	2A2	2A4	MAYBE INTO 2A1
46-49	49-46	DS04654	2B2	2B9	MAYBE TO 2B1 AND/OR 2B10
53-2	2-53	DS00149	2B8	2B9	MAYBE TO 2B7 &/OR 2B10
45-48	48-45	DS04557	2B8	2B9	MAYBE TO 2B7 &/OR 2B10
90-77	77-89	DS07385	2B11	2B15	
136-60	60-93	DS05757	2B11	2B18	
50-67	67-50	DS06386	2C2	2C4,5	MAYBE TO 2C1
139-81	81-96	DS07776	2C8	2D1	MAYBE TO 2C7 AND/OR 2D2
81-19	Not present	DS09235	2F1	2F2	
131-11	Not present	DS09323	2F1	2F5	
41-34	34-41	DS03209	2F1	2F6	
50-32	32-50	DS03026	2F3	2F6	
25-75	75-25	DS07129	2F3	2F6	ALSO SECONDARY SITES IN SAME REGION=DS07225
25-76	76-25	DS07225	2F3	2F6	ALSO SECONDARY SITES IN SAME REGION=DS07129
49-32	32-49	DS03025	2F4	2F5	
25-38	38-25	DS03577	3A1	3A2	
41-11	11-41	DS01001	3A1	3A4	
65-65	65-65	DS06209	3A3	3A4	
46-75	75-46	DS07150	3A3	3A4	
41-27	27-41	DS02537	3A5	3A10	
137-54	54-94	DS05182	3A5	3A10	ALSO CC LIGHT + ~5-10 EUCHROMATIC SITES
135-55	55-92	DS05276	3A6		
45-25	25-45	DS02349	3A6	3B2	ALSO ~5 OTHER SITES & CHROMOCENTER
42-58	58-42	DS05514	3A6	3B2	CC LIGHT & ~10 SECONDARY SITES
136-50	50-93	DS04797	3A8	3B2	
	5-58	DS00442	3A8	3B4	
24-38	38-24	DS03576	3C1	3C2	MAYBE TO 3C3
135-71	71-92	DS06812	3C1	3C2	MAYBE TO 3C3
138-66	66-95	DS06335	3C1	3C5	MAYBE TO 3C6

CAMBR	BERK	DS No.	BEGIN	END	COMMENTS
24-96	96-24	DS09144	3C3	3C5	MAYBE TO 3C2 &/OR 3C6 + CC LIGHT + ~200 SITES
42-33	33-42	DS03114	3C3	3C7	ALSO ~100 SECONDARY SITES
49-69	69-49	DS06577	3C7		
	4-89	DS00377	3C9	3C10	ALSO CC LIGHT + ~150 SECONDARY SITES
	7-22	DS00598	3D3	3D4	
62-45	45-62	DS04286	3D6	3E1	MAYBE TO 3D5 &/OR 3E2; ALSO CC LIGHT + ~100 SITES
64-75	75-64	DS07168	3E6	3F1	MAYBE TO 3F2; SAME AS DS07264
64-76	76-64	DS07264	3E6	3F1	MAYBE TO 3F2; SAME AS DS07168
64-20	20-64	DS01888	3F1	3F2	
62-87	87-62	DS08318	3F1	3F2	SAME AS DS08702
62-91	91-62	DS08702	3F1	3F2	SAME AS DS08318
	1-32	DS00032	3F2	3F3	MAYBE TO 3F1
43-40	40-43	DS03787	3F3		
76-1	1-76	DS00076	3F4,5		
	6-23	DS00503	3F4,5		
43-15	15-43	DS01387	3F4,5		
46-27	27-46	DS02542	4A1	4A2	
139-47	47-96	DS04512	4A1	4A2	SAME AS DS05088
139-53	53-96	DS05088	4A1	4A2	SAME AS DS04512
131-54	Not present	DS09366	4A1	4A2	CC LIGHT + ~25 SECONDARY SITES
	1-36	DS00036	4A3	4A5	MAYBE TO 4A6
50-2	2-50	DS00146	4A4	4B1	MAYBE TO 4B2; =DS01586
50-17	17-50	DS01586	4A4	4B1	MAYBE TO 4B2; =DS00146
	6-60	DS00540	4B4	4C1	MAYBE TO 4B3 &/OR 4C2 + CC LIGHT
47-27	27-47	DS02543	4B4	4C1	MAYBE 4B3 &/OR 4C2 + CC LITE + ~50 SCNDRY SITES
131-96	Not present	DS09408	4B5	4B6	
	4-16	DS00304	4C1	4C2	
65-2	2-65	DS00161	4C2	4C3	MAYBE TO 4C1 &/OR 4C4
90-62	62-89	DS05945	4C2	4C3	MAYBE TO 4C1 &/OR 4C4
62-44	44-62	DS04190	4C3	4C4	
	8-38	DS00710	4C7	4C8	
62-27	27-62	DS02558	4C7	4C8	
24-27	27-24	DS02520	4D3	4E1	MAYBE TO 4E2
	5-73	DS00457	4E1	4E2	ALSO SOME WEAK SECONDARY SITES
63-82	82-63	DS07839	4E1	4E2	
	5-14	DS00398	4F2	4F3	MAYBE TO 4F1 &/OR 4F4
63-59	59-63	DS05631	4F2	4F3	MAYBE TO 4F1 &/OR 4F4
47-95	95-47	DS09071	4F2	4F3	MAYBE TO 4F2 &/OR 4F4
135-6	6-92	DS00572	4F11,12	5A1	MAYBE TO 5A2
49-45	45-49	DS04273	4F11,12	5A1	MAYBE TO 5A2
65-76	76-65	DS07265	5A9	5B1	MAYBE TO 5A8 &/OR 5B2; SAME AS DS07361
65-77	77-65	DS07361	5A9	5B1	MAYBE TO 5A8 &/OR 5B2; SAME AS DS07265
	7-13	DS00589	5C1	5C2	
50-75	75-50	DS07154	5C1	5C2	
	5-84	DS00468	5C3	5C5	MAYBE TO 5C6
25-67	67-25	DS06361	5D1	5D2	
131-53	Not present	DS09365	5D1	5D2	
25-87	87-25	DS08281	5D2	5D5	MAYBE TO 5D1 &/OR 5D6
	5-38	DS00422	5D3,4		
47-36	36-47	DS03407	5D3,4	5D5	MAYBE TO 5D6 + CC MODERATE + ~10 SECONDARY SITES
	8-43	DS00715	5E1,2	5E4	MAYBE TO 5E5
	4-87	DS00375	5E5	5F1	MAYBE TO 5F2
135-70	70-92	DS06716	5E5	5F6	MAYBE TO 5E4

CAMBR	BERK	DS No.	BEGIN	END	COMMENTS
61-3	3-61	DS00253	5F1	5F5	
	4-93	DS00381	6A1	6A2	ALSO CC LIGHT + MANY WEAK SITES
	6-13	DS00493	6A2	6B1	MAYBE TO 6A1 &/OR 6B2
137-89	89-94	DS08542	6A2	6B1	MAYBE TO 6A1 &/OR 6B2
	2-23	DS00119	6A4	6B1	MAYBE TO 6A3 &/OR 6B2; ALSO CC LIGHT + ~ 100 SITES
	9-3	DS00771	6B2	6C1	MAYBE TO 6B1 &/OR 6C2
138-52	52-95	DS04991	6B2	6C1	MAYBE TO 6B1 &/OR 6C2 + CC LITE + ~25 MORE SITES
90-66	66-89	DS06329	6C1	6C2	
24-47	47-24	DS04440	6E1	6E2	
	6-85	DS00565	6F1,2	6F5,6	
	8-73	DS00745	6F1,2	6F5,6	
	8-53	DS00725	6F1,2		
25-50	50-25	DS04729	7A2	7A4,5	MAYBE TO 07A1
	9-73	DS00841	7A4,5		
136-58	58-93	DS05565	7A6	7A8	
135-2	2-92	DS00188	7B3	7B6	
	9-60	DS00828	7B3	7B8	
136-75	75-93	DS07197	7C1	7C2	
	4-45	DS00333	7D1	7D2	
41-16	16-41	DS01481	7D1	7D2	
43-86	86-43	DS08203	7D1	7D2	
24-21	21-24	DS01944	7D6,7	7D16	MAYBE TO 7D17
50-88	88-50	DS08402	7D10	7D18,19	
81-34	Not present	DS09250	7D14	7D22	
44-53	53-44	DS05036	7E2	7E5	MAYBE TO 7E1
65-43	43-65	DS04097	7E2	7E6	MAYBE TO 7E1
46-52	52-46	DS04942	8A1	8A2	
136-94	94-93	DS09021	8B5	8B8	
50-77	77-50	DS07346	8C1,2		
	5-75	DS00459	8C9,10	8C13	MAYBE TO 8C14
131-30	Not present	DS09342	8D2	8D4	MAYBE TO 8D1 &/OR 8D5
90-64	64-89	DS06137	8D10,11	8D12	
64-91	91-64	DS08704	8D10,11	8D12	ALSO 23C1;23C2 LIGHT
	5-89	DS00473	8E1	8E2	
48-21	21-48	DS01968	8E3	8E4	
135-57	57-92	DS05468	8E3	8E4	
41-57	57-41	DS05417	8F1	8F2	
47-15	15-47	DS01391	9A1	9A2	SAME AS DS00719 AND DS01007
	8-47	DS00719	9A1	9A3	SAME AS DS01391 AND DS01007
47-11	11-47	DS01007	9A1	9A4	SAME AS DS00719 AND DS01391
42-76	76-42	DS07242	9A1	9A4	
42-63	63-42	DS05994	9A1	9A5	CC LIGHT & ~ 25 SECONDARY SITES
71-1	1-71	DS00071	9A2	9A3	
72-1	1-72	DS00072	9A2	9A3	ALSO ~15 LIGHTER SITES
	33-14	DS03086	9A3	9A4	
	74-91	DS07099	9A3	9A4	
50-34	34-50	DS03218	9A4		
48-61	61-48	DS05808	9A4		
137-30	30-94	DS02878	9B1	9B2	ALSO CC LIGHT + ~25 SECONDARY SITES
49-34	34-49	DS03217	9B1	9B2	
	6-3	DS00483	9B12	9B14,15	MAYBE TO 9B11
63-37	37-63	DS03519	9B14,15	9C1	MAYBE TO 9C2 + ~10 SECONDARY SITES
62-71	71-62	DS06782	9B14,15	9C1	MAYBE TO 9C2 + ~25 WEAK SECONDARY SITES

CAMBR	BERK	DS No.	BEGIN	END	COMMENTS
25-73	73-25	DS06937	9C1	9C2	
81-29	Not present	DS09245	9D1	9D2	
131-80	Not present	DS09392	9D1	9D2	
81-89	Not present	DS09305	9D3	9D4	
46-65	65-46	DS06190	9E2	9F5	MAYBE TO 9E1 AND/OR 9F6
63-38	38-63	DS03615	9E7,8	9F5,6	
62-47	47-62	DS04478	9F1	9F2	
	1-18	DS00018	9F11	9F13	MAYBE TO 9F10
43-10	10-43	DS00907	10A1	10A2	
62-14	14-62	DS01310	10A1	10A2	SAME AS DS01790
62-19	19-62	DS01790	10A1	10A2	SAME AS DS01310
136-20	20-93	DS01917	10A1	10A2	ALSO ~25 WEAK SECONDARY SITES
90-22	22-89	DS02105	10A1	10A2	
139-26	26-96	DS02496	10A1	10A2	
81-72	Not present	DS09288	10A1	10A2	
136-86	86-93	DS08253	10A1		
	8-34	DS00706	10B1	10B2	ALSO SOME SECONDARY SITES
137-82	82-94	DS07870	10B3	10B10	MAYBE TO 10B11 + CC LIGHT
	6-70	DS00550	10B15,16	10C1	MAYBE TO 10C2 + CC LIGHT + SOME LIGHT SITES
136-83	83-93	DS07965	10C1	10C2	
	7-93	DS00669	10D3	10D5	POSSIBLY SAME AS DS00668
135-7	7-92	DS00668	10D3	10D6	
15-17	17-15	DS01551	10D5	10E3,4	cDNA 3b6
9-48	48-9	DS04521	10D5	10E3,4	cDNA 3b6
137-77	77-94	DS07390	10E5	10F11	
138-16	16-95	DS01535	10F2	10F11	MAYBE TO 10F1
138-65	65-95	DS06239	10F3	10F11	
25-21	21-25	DS01945	11A7	11A8	MAYBE TO 11A6 &/OR 11A9
63-91	91-63	DS08703	11A7	11A8	MAYBE TO 11A6 &/OR 11A9 + ~25 WEAKER SITES
43-31	31-43	DS02923	11A8		
81-77	Not present	DS09293	11A8		
63-27	27-63	DS02559	11A10	11A11	MAYBE TO 11A12
139-94	94-96	DS09024	11A10	11A11	MAYBE TO 11A12
64-33	33-64	DS03136	11B1	11B2	
139-92	92-96	DS08832	11B3	11B9	
	9-61	DS00829	11C1,2		
135-86	86-92	DS08252	11C1,2		
81-87	Not present	DS09303	11C1,2		
24-1	1-24	DS00024	11D8	11E1,2	
	8-8	DS00680	11D9	11D10	
47-33	33-47	DS03119	11E9,10	11F1,2	
90-54	54-89	DS05177	12A1	12A2	
42-92	92-42	DS08778	12A1	12A4,5	
62-38	38-62	DS03614	12A3	12A10	
46-37	37-46	DS03502	12B1	12B2	
44-60	60-44	DS05708	12B1	12B2	
135-78	78-92	DS07484	12B1	12B2	
26-4	4-26	DS00314	12D1	12D2	cDNA 4c9
	6-67	DS00547	12E1	12E2	
25-26	26-25	DS02425	12E1	12E2	CC LIGHT + ~100 SECONDARY SITES
64-53	53-64	DS05056	12E1	12E2	
43-1	1-43	DS00043	12E3	12E7	ALSO CC LIGHT
138-58	58-95	DS05567	12E3	12E7	

CAMBR	BERK	DS No.	BEGIN	END	COMMENTS
135-63	63-92	DS06044	12E3	12E7	
138-57	57-95	DS05471	12E8	12E9	ALSO ~10 SECONDARY EUCHROM SITES
90-46	46-89	DS04409	12E9	12F1,2	MAYBE TO 12E8 & ~25 SECONDARY SITES
44-63	63-44	DS05996	12E10	12E11	
	6-11	DS00491	12F1,2		
62-22	22-62	DS02078	12F1,2		
24-73	73-24	DS06936	12F1,2		CC LIGHT + SOME LIGHT EUCHROMATIC SITES
	9-75	DS00843	12F4	12F5	
131-73	Not present	DS09385	13A1,2	13A5,6	
	4-63	DS00351	13A1,2		
41-69	69-41	DS06569	13A8,9		
48-33	33-48	DS03120	13B1,2	13B3	MAYBE TO 13B4 + CC LIGHT + ~5-10 WK OTHER SITES
48-45	45-48	DS04272	13B1,2	13B3	MAYBE TO 13B4
138-89	89-95	DS08543	13B3	13B4	
24-29	29-24	DS02712	13B5	13C1,2	
	3-29	DS00221	13C1,2		
	7-19	DS00595	13C1,2		
	1-37	DS00037	13E12,13	13F1,2	
138-83	83-95	DS07967	13E14,15	13F1,2	
43-96	96-43	DS09163	13E14,15		
	5-10	DS00394	13F1,2	13F10,11	
90-38	38-89	DS03641	14A3	14A5	
60-3	3-60	DS00252	14B1,2	14B3,4	ALSO CC LIGHT + SOME WEAK SCONDARY SITES
	5-63	DS00447	14B1,2	14B3,4	ALSO ~100 SECONDARY SITES
	4-47	DS00335	14B3,4		
25-58	58-25	DS05497	14B3,4		SOME VERY LIGHT SECONDARY SITES
46-18	18-46	DS01678	14B14	14C1,2	MAYBE TO 14B13
49-23	23-49	DS02161	14C1,2	14C4,5	
	9-18	DS00786	14C1,2		
	68-8	DS06440	14C3	14C8	
	53-93	DS05085	14D1,2		
139-50	50-96	DS04800	14E1,2	14F2,3	
46-74	74-46	DS07054	14F6	15A3,4	MAYBE TO 14F5
64-55	55-64	DS05248	15A1,2	15A3,4	
135-94	94-92	DS09020	15A1,2	15A3,4	
45-45	45-45	DS04269	15F6	15F9	
	3-73	DS00265	16A1,2	16A5,6	
	3-79	DS00271	16A1,2		
	5-9	DS00393	16A1,2		
47-24	24-47	DS02255	16A1,2		
50-71	71-50	DS06770	16A1,2		
90-59	59-89	DS05657	16A3?	16A5,6	THE ? IS SORSA'S
	4-21	DS00309	16B1,2	16B4,5	
	9-76	DS00844	16B1,2		
	3-34	DS00226	16D1	16D2	
64-58	58-64	DS05536	16D2	16D6	MAYBE TO 16D1
65-39	39-65	DS03713	16F1,2	16F3,4	
65-48	48-65	DS04577	16F1,2	16F3,4	
63-63	63-63	DS06015	16F1,2	16F3,4	
48-32	32-48	DS03024	16F1,2	16F7,8	PROBABLY SAME CLONE AS DS01104
131-95	Not present	DS09407	16F1,2	16F7,8	
48-12	12-48	DS01104	16F3,4	16F7,8	PROBABLY SAME CLONE AS DS03024
48-80	80-48	DS07632	16F5	16F7,8	
49-82	82-49	DS07825	16F7,8	17A1	MAYBE TO 17A2

CAMBR	BERK	DS No.	BEGIN	END	COMMENTS
	6-10	DS00490	17A1	17A2	
50-82	82-50	DS07826	17A1	17A2	
24-25	25-24	DS02328	17A2	17A3	MAYBE TO 17A1 &/0R 17A4
47-56	56-47	DS05327	17A2	17A3	MAYBE TO 17A1 &/OR 17A4
	8-78	DS00750	17A7	17B1,2	
	6-6	DS00486	17A11,12	17B1,2	ALSO SOME SECONDARY SITES
	6-2	DS00482	17B1,2		ALSO SOME WEAK SECONDARY SITES
	8-77	DS00749	17B5	17C3,4	MAYBE TO 17B4
64-78	78-64	DS07456	17D1,2	17D6	
136-15	15-93	DS01437	18A2	18A3,4	MAYBE TO 18A1 + CC LIGHT + ~25 SECONDARY SITES
138-50	50-95	DS04799	18A2	18A3,4	MAYBE TO 18A1 + CC MODERATE + ~150 MORE SITES
137-1	1-94	DS00094	18A3,4		ALSO ~8-10 WEAK SECONDARY SITES
69-2	2-69	DS00165	18A3,4		ALSO ~20 WEAK SITES
81-96	96-81	DS09201	18A3,4		
62-76	76-62	DS07262	18B1,2	18B7	
65-38	38-65	DS03617	18B1,2	18B8?	QUOTATION MARK IS SORSA'S DESIGNATION
44-91	91-44	DS08684	18C2	18C3,4	MAYBE TO 18C1; SAME AS DS08780
44-92	92-44	DS08780	18C2	18C3,4	MAYBE TO 18C1; DOTS ON CC; SAME AS DS08684
	7-35	DS00611	18C8,9	18D3,4	
42-43	43-42	DS04074	18D1,2		
43-61	61-43	DS05803	18F1	18F2	
65-30	30-65	DS02849	18F3	18F5	
	8-76	DS00748	19A1,2	19A3	MAYBE TO 19A4
135-37	37-92	DS03548	19A1,2	19A3	MAYBE TO 19A4
135-88	88-92	DS08444	19C1,2		
41-90	90-41	DS08585	19C1,2		
42-38	38-42	DS03594	19D1	19D3	
62-25	25-62	DS02366	19E1	19E2	
138-28	28-95	DS02687	19E1	19E2	
63-51	51-63	DS04863	19E1	19E2	
	6-72	DS00552	19E2	19E3	MAYBE TO 19E1 &/OR 19E4
62-16	16-62	DS01502	19E2	19E3	MAYBE TO 19E1 &/OR 19E4
43-24	24-43	DS02251	19E2	19E3	MAYBE TO 19E1 AND/OR 19E4
46-46	46-46	DS04366	19E2	19E3	MAYBE TO 19E1 AND/OR 19E4
137-83	83-94	DS07966	19E2	19E3	MAYBE TO 19E1 &/OR 19E4
138-91	91-95	DS08735	19E2	19E3	MAYBE TO 19E1 &/OR 19E4
	5-60	DS00444	19E3	19E4	
	7-15	DS00591	19E3	19E4	
	9-85	DS00853	19E3	19E4	
63-16	16-63	DS01503	19E3	19E4	
24-43	43-24	DS04056	19E3	19E4	
64-87	87-64	DS08320	19E4	19E6	MAYBE TO 19E3
137-19	19-94	DS01822	19E7,8		
55-3	3-55	DS00247	19F1	19F2	
63-28	28-63	DS02655	19F1	19F2	CC VERY LIGHT
49-43	43-49	DS04081	19F1	19F2	
65-67	67-65	DS06401	19F1	19F2	
42-68	68-42	DS06474	19F1	19F2	
	4-61	DS00349	20A1,2		
65-31	31-65	DS02945	20A1,2		ALSO ~75 WEAK SECONDARY SITES
47-66	66-47	DS06287	20A1,2		ALSO CC LIGHT; PROBABLY SAME CLONE AS DS06959
47-73	73-47	DS06959	20A1,2		PROBABLY SAME CLONE AS DS06287

CAMBR	BERK	DS No.	BEGIN	END	COMMENTS
49-90	90-49	DS08593	20A3		
138-45	45-95	DS04319	20A4	20A5	
	7-95	DS00671	20B1,2		ALSO CC LIGHT + ~10 SECONDARY SITES
50-22	22-50	DS02066	20B1,2		ALSO CC VERY LIGHT
90-70	70-89	DS06713	20B1,2		
	8-60	DS00732	20C3		
43-33	33-43	DS03115	21A1	21A3	NEAR OR AT TELOMERE & ~10 LIGHT SITES
	3-82	DS00274	21A1	21A4	ALSO ~10 LIGHT SITES
	9-19	DS00787	21A1	21A4	
64-26	26-64	DS02464	21A1	21A5	ALSO SOME WEAK SITES
135-46	46-92	DS04412	21A1		ALSO ~15 SECONDARY SITES
	8-66	DS00738	21B1	21B4,5	
41-74	74-41	DS07049	21B2	21B8	
81-69	Not present	DS09285	21B3	21B7	
83-2	2-83	DS00179	21B4,5	21B8	
81-75	Not present	DS09291	21B4,5		
	4-59	DS00347	21B6	21B8	
	8-7	DS00679	21C1	21C2	
137-48	48-94	DS04606	21C1	21C2	
	9-39	DS00807	21C2	21C4	MAYBE TO 21C1
135-8	8-92	DS00764	21C3	21C4	
62-53	53-62	DS05054	21C4		ALSO ~50 WEAK SECONDARY SITES
	6-21	DS00501	21C4	21C7	
41-47	47-41	DS04457	21C6	21C8	
135-20	20-92	DS01916	21D1		
	3-77	DS00269	21D1	21D2	
	4-29	DS00317	21D1	21D2	
25-60	60-25	DS05689	21D1	21D2	
139-76	76-96	DS07296	21D1	21D2	
25-72	72-25	DS06841	21D3	21D4	
135-28	28-92	DS02684	21D3	21E1	MAYBE TO 21E2
136-67	67-93	DS06429	21D3	21E1	
65-12	12-65	DS01121	21E1	21E2	ALSO ~ 100 SECONDARY SITES
43-16	16-43	DS01483	21E1	21E2	
65-55	55-65	DS05249	21E1	21E2	
135-83	83-92	DS07964	21E1	21E2	ALSO ~50 VERY WEAK SECONDARY SITES
138-32	32-95	DS03071	21E3	21E4	
	7-40	DS00616	21E3	21F3	
49-38	38-49	DS03601	22A1		
50-38	38-50	DS03602	22A1		
	3-85	DS00277	22A1	22A2	
	4-41	DS00329	22A1	22A2	
41-26	26-41	DS02441	22A1	22A2	
48-29	29-48	DS02736	22A1	22A2	
135-51	51-92	DS04892	22A1	22A2	
41-82	82-41	DS07817	22A1	22A2	ALSO LIGHT CC & ~10 WEAK SITES
42-82	82-42	DS07818	22A1	22A2	
62-56	56-62	DS05342	22A2	22A3	MAYBE TO 22A1; ALSO CC LIGHT + SOME SECNDRY SITES
	8-4	DS00676	22A3		ALSO CC LIGHT + SOME LIGHT SECONDARY SITES
136-65	65-93	DS06237	22A3		CC LIGHT + ~5-10 WEAK SECONDARY SITES
62-80	80-62	DS07646	22A3		
81-30	Not present	DS09246	22A3		ALSO ~50 SECONDARY SITES
	5-61	DS00445	22A3	22A5	

CAMBR	BERK	DS No.	BEGIN	END	COMMENTS
63-55	55-63	DS05247	22A4	22B1	MAYBE TO 22B2
42-48	48-42	DS04554	22A5	22A8	SOME SECONDARY SITES WHEN LABEL IS HEAVY
50-10	10-50	DS00914	22B1	22B2	
62-59	59-62	DS05630	22B1	22B2	
131-60	Not present	DS09372	22B1	22B2	
42-67	67-42	DS06378	22B3	22B9	
68-2	2-68	DS00164	22C1,2		ALSO ~25 WEAK SITES
64-19	19-64	DS01792	22C1,2		ALSO ~10 WEAK SECONDARY SITES
64-83	83-64	DS07936	22C1,2		ALSO ~100 SECONDARY SITES
42-85	85-42	DS08106	22C1,2		
131-88	Not present	DS09400	22D1+	22D6	22D1+ DESIGNATION IS SORSA'S
	4-62	DS00350	22E1,2		
	5-18	DS00402	22E1,2		
65-25	25-65	DS02369	22E1,2		
138-72	72-95	DS06911	22E1,2		ALSO ~25 SECONDARY EUCHROM SITES
81-39	Not present	DS09255	22E1,2		ALSO ~10 SECONDARY SITES
42-4	4-42	DS00330	22F1	22F2	
	5-53	DS00437	22F1	22F2	
137-13	13-94	DS01246	22F1	22F2	
43-21	21-43	DS01963	22F1	22F2	
63-73	73-63	DS06975	22F1	22F2	
44-95	95-44	DS09068	22F1	22F2	
	7-60	DS00636	22F2	22F3+	22F3+ DESIG SORSA; MAYBE TO 22F1; ALSO 72E3 LIGHT
135-66	66-92	DS06332	22F2	22F4	MAYBE TO 22F1 + ~25 SECONDARY SITES
46-21	21-46	DS01966	22F3+	22F4	22F3+ IS SORSA'S DESIGNATION
138-96	96-95	DS09215	22F3+	22F4	22F3+ IS SORSA'S DESIGNATION FOR THIS BAND
	3-76	DS00268	22F3+		BAND DESIGNATION IS SORSA'S
52-3	3-52	DS00244	23A1	23A2	
65-11	11-65	DS01025	23A1	23A2	
50-14	14-50	DS01298	23A1	23A2	
137-29	29-94	DS02782	23A1	23A2	ALSO ~25 SECONDARY SITES
64-67	67-64	DS06400	23A1	23A2	
43-94	94-43	DS08971	23A1	23A2	
64-96	96-64	DS09184	23A1	23A2	
45-75	75-45	DS07149	23B2	23C1	MAYBE TO 23B1 &/OR 23C2
42-10	10-42	DS00906	23C1	23C2	
135-60	60-92	DS05756	23C1	23C2	
	8-12	DS00684	23C3	23C5	
47-89	89-47	DS08495	23D1	23D2	ALSO ~15 SECONDARY SITES
45-89	89-45	DS08493	23D2	23D4	MAYBE TO 23D1 & SOME LIGHT HYBRIDIZATION IN CC
25-95	95-25	DS09049	23D2	23D6	MAYBE TO 23D1
	2-10	DS00106	23E1,2	23E4	ALSO ~50 SECONDARY SITES
24-22	22-24	DS02040	23E1,2	23E4	
50-50	50-50	DS04754	23E1,2	23E4	
135-75	75-92	DS07196	23E1,2	23E4	
51-1	1-51	DS00051	23E1,2		ALSO SOME LIGHT SECONDARY SITES
65-26	26-65	DS02465	23F1,2	23F6	
	4-36	DS00324	23F1,2		
135-49	49-92	DS04700	23F1,2		
135-14	14-92	DS01340	24A1	24A2	
131-32	Not present	DS09344	24A1	24A2	
131-37	Not present	DS09349	24A1	24A2	
24-3	3-24	DS00216	24A3	24B1,2	

CAMBR	BERK	DS No.	BEGIN	END	COMMENTS
63-62	62-63	DS05919	24B1,2	24C1,2	
65-84	84-65	DS08033	24B1,2	24C1,2	
137-24	24-94	DS02302	24C1,2	24C5	
65-1	1-65	DS00065	24C3	24C9	
62-42	42-62	DS03998	24D1	24D2	
81-28	Not present	DS09244	24D3		
62-50	50-62	DS04766	24D3	24D5	MAYBE TO 24D6
	8-33	DS00705	24D3	24D8	
50-68	68-50	DS06482	24D5	24D6	
131-9	Not present	DS09321	24D5	24D6	
	2-26	DS00122	24E1,2		
	2-34	DS00130	24E1,2		
	3-6	DS00198	24E1,2		
64-4	4-64	DS00352	24E1,2		
62-9	9-62	DS00830	24E1,2		
42-80	80-42	DS07626	24E1,2		
131-48	Not present	DS09360	24E1,2		
64-44	44-64	DS04192	24E1,2	24F1	MAYBE TO 24F2
90-55	55-89	DS05273	24F1	24F2	
139-83	83-96	DS07968	24F2	24F7	MAYBE TO 24F1
41-31	31-41	DS02921	24F3	25A1	MAYBE TO 25A2
	3-9	DS00201	25A1	25A2	ALSO ~10 SECONDARY SITES
	3-86	DS00278	25A1	25A2	
49-59	59-49	DS05617	25A1	25A2	
45-82	82-45	DS07821	25A2	25A3	MAYBE TO 25A1 &/OR 25A4
42-1	1-42	DS00042	25A3	25A4	
84-1	1-84	DS00084	25A3	25A4	
43-42	42-43	DS03979	25C5	25C6	
24-14	14-24	DS01272	25C6	25C10	MAYBE TO 25C5
47-70	70-47	DS06671	25C6	25D1	MAYBE TO 25C5 &/OR 25D2
46-80	80-46	DS07630	25D1	25D2	
42-8	8-42	DS00714	25D4	25D7	LITE CC & <10 EUCHR SITES
136-71	71-93	DS06813	25D4	25D7	CC LIGHT + ~50 WEAK SECONDARY SITES
44-35	35-44	DS03308	25E6		
	4-27	DS00315	25F1	25F2	
	8-83	DS00755	25F1	25F2	
63-86	86-63	DS08223	25F2	25F3	MAYBE TO 25F1 &/OR 25F4
64-23	23-64	DS02176	25F3	25F4	ALSO CC LIGHT + ~10 WEAK SITES
45-12	12-45	DS01101	25F4	25F5	MAYBE TO 25F3
48-91	91-48	DS08688	25F4	25F5	MAYBE TO 25F3
	2-11	DS00107	26A1	26A2	
72-2	2-72	DS00168	26A1	26A2	
25-34	34-25	DS03193	26A1	26A2	
47-57	57-47	DS05423	26A1	26A2	ALSO CC LIGHT + ~150 SECONDARY SITES
45-68	68-45	DS06477	26A1	26A2	
138-21	21-95	DS02015	26A5	26A6	
42-88	88-42	DS08394	26A5	26A6	
138-31	31-95	DS02975	26B1,2	26B5,6	
24-23	23-24	DS02136	26B5,6	26B11	
47-72	72-47	DS06863	26B9	26C1	MAYBE TO 26B8 &/OR 26C2
50-53	53-50	DS05042	26C1		
	2-25	DS00121	26C1	26C2	
138-18	18-95	DS01727	26C1	26C2	

CAMBR	BERK	DS No.	BEGIN	END	COMMENTS
41-45	45-41	DS04265	26C1	26C2	
43-52	52-43	DS04939	26C1	26C2	
137-57	57-94	DS05470	26C1	26C2	
43-73	73-43	DS06955	26C1	26C2	
131-18	Not present	DS09330	26C1	26C2	
	1-23	DS00023	26C3	26C4	
50-49	49-50	DS04658	26C3	26C4	
62-21	21-62	DS01982	26C2	26D1,2	MAYBE TO 26C1
49-24	24-49	DS02257	26C2	26D1,2	MAYBE TO 26C1 + ~6 WEAK SECONDARY SITES
	8-28	DS00700	26C3	26D1,2	
50-24	24-50	DS02258	26C3	26D1,2	
135-1	1-92	DS00092	26D1,2		ALSO ~100 SECONDARY SITES
	2-12	DS00108	26D1,2		
46-7	7-46	DS00622	26D1,2		
41-20	20-41	DS01865	26D1,2		
81-51	Not present	DS09267	26D1,2		ALSO ~10 WEAK SITES
131-34	Not present	DS09346	26D1,2		ALSO ~100 SECONDARY SITES
81-2	Not present	DS09218	26D3	26D11	
49-48	48-49	DS04561	26D4	26D10	SIGNAL INTERBAND-LIKE ACROSS WHOLE REGION
	7-71	DS00647	26F1,2	26F3,4	
136-81	81-93	DS07773	26F1,2	26F3,4	
65-87	87-65	DS08321	26F1,2	26F3,4	
	5-45	DS00429	26F3,4	26F7	
50-41	41-50	DS03890	26F3,4	27A1	
64-31	31-64	DS02944	27A1	27B1,2	
90-72	72-89	DS06905	27A2	27B1,2	
81-12	Not present	DS09228	27A2	27C1	MAYBE TO 27A1 &/OR 27C2
43-68	68-43	DS06475	27B1,2	27B4	
	5-7	DS00391	27B1,2	27C1	
43-30	30-43	DS02827	27B1,2	27C1	
24-31	31-24	DS02904	27B4	27C1	
48-50	50-48	DS04752	27C1		SAME AS DS08112
48-85	85-48	DS08112	27C1		SAME AS DS04752
45-57	57-45	DS05421	27C1	27C2	
	4-67	DS00355	27C4,5	27D1,2	
137-68	68-94	DS06526	27E2	27E8?	MAYBE TO 27E1;27E8? IS SORSA'S DESIGNATION
47-26	26-47	DS02447	27E3	27F1,2	
2-67	67-2	DS06338	27F1,2		
90-95	95-89	DS09113	27F1,2		
131-83	Not present	DS09395	27F6	27F7	
	2-4	DS00100	28A3	28A6	
90-16	16-89	DS01529	28B1,2		
49-84	84-49	DS08017	28B1,2		
42-75	75-42	DS07146	28B1,2	28B4	
	4-1	DS00289	28C4,5	28D2	
	4-85	DS00373	28D1	28D3,4	
131-23	Not present	DS09335	28D1	28D3,4	
45-90	90-45	DS08589	28E1		
24-28	28-24	DS02616	28E1	28E3	MAYBE TO 28E4
25-51	51-25	DS04825	28E3	28E4	
44-66	66-44	DS06284	28E3	28E4	
50-5	5-50	DS00434	29A1	29A2	
90-57	57-89	DS05465	29A2	29B1	MAYBE TO 29A1 &/OR 29B2

CAMBR	*BERK*	*DS No.*	*BEGIN*	*END*	*COMMENTS*
139-89	89-96	DS08544	29A2	29B1	MAYBE TO 29A1 AND/OR 29B2
136-90	90-93	DS08637	29B2	29B4	MAYBE TO 29B1
24-67	67-24	DS06360	29B2	29C3	MAYBE TO 29B1
24-85	85-24	DS08088	29C1,2	29C4	
137-22	22-94	DS02110	29C1,2	29C5	
43-35	35-43	DS03307	29D1,2		
43-38	38-43	DS03595	29D1,2		
50-92	92-50	DS08786	29E1	29E2	ALSO ~25 SECONDARY SITES
	7-18	DS00594	29E4	29F1	MAYBE TO 29E3 &/OR 29F2
24-92	92-24	DS08760	29F2	29F3	MAYBE TO 29F1
54-1	1-54	DS00054	30A1	30A2	
137-11	11-94	DS01054	30A1	30A2	
63-19	19-63	DS01791	30A1	30A2	
42-32	32-42	DS03018	30A3	30A6	
25-33	33-25	DS03097	30A3	30A6	CC LIGHT + ~25 SECONDARY SITES
25-57	57-25	DS05401	30A3	30A6	CC VERY LIGHT
48-60	60-48	DS05712	30A3	30A6	ALSO CC LIGHT + FEW LIGHT SECONDARY SITES
46-73	73-46	DS06958	30A3	30A6	ALSO CC LIGHT
49-40	40-49	DS03793	30A4	30A6	
137-21	21-94	DS02014	30A7	30A8	
46-36	36-46	DS03406	30A7	30A8	
65-40	40-65	DS03809	30A7	30A8	
63-65	65-63	DS06207	30A7	30A8	
46-68	68-46	DS06478	30A7	30A8	
63-79	79-63	DS07551	30A8	30B1,2	MAYBE TO 30A7
48-67	67-48	DS06384	30B5,6	30B9	MAYBE TO 30B10
45-35	35-45	DS03309	30B9	30B10	
41-49	49-41	DS04649	30B9	30B10	
81-47	Not present	DS09263	30C1	30C2	
24-40	40-24	DS03768	30C2	30C5	MAYBE TO 30C1
	1-28	DS00028	30C5	30D1,2	
	5-76	DS00460	30D1,2		
	9-74	DS00842	30D1,2		
46-42	42-46	DS03982	30D1,2		
49-46	46-49	DS04369	30D1,2		
90-87	87-89	DS08345	30D1,2		
62-74	74-62	DS07070	30D1,2	30E1,2	
131-87	Not present	DS09399	30D3	30E1,2	
58-1	1-58	DS00058	31A1	31A2	
24-10	10-24	DS00888	31A1	31A2	
49-47	47-49	DS04465	31A1	31A2	
41-61	61-41	DS05801	31A1	31A2	
49-76	76-49	DS07249	31B1		
47-92	92-47	DS08783	31C1,2	31D6	SAME AS DS08879
47-93	93-47	DS08879	31C1,2	31D6	SAME AS DS08783
	9-27	DS00795	31D9	31E1	MAYBE TO 31D8
90-10	10-89	DS00953	31F1,2		
139-31	31-96	DS02976	31F3	31F5	
42-24	24-42	DS02250	32A1		
	5-93	DS00477	32A1	32A2	ALSO ~50 SECONDARY SITES
	8-22	DS00694	32A1	32A2	
63-24	24-63	DS02271	32A1	32A2	ALSO ~25 MODERATELY INTENSE SECONDARY SITES
44-81	81-44	DS07724	32A1	32A2	

CAMBR	BERK	DS No.	BEGIN	END	COMMENTS
81-71	Not present	DS09287	32A1	32A2	
24-65	65-24	DS06168	32A2	32A3	MAYBE TO 32A1 &/OR 32A4
49-88	88-49	DS08401	32A3	32A4	
44-44	44-44	DS04172	32A3	32B4	
64-65	65-64	DS06208	32B1,2		
131-52	Not present	DS09364	32B1,2		
	1-34	DS00034	32C1,2		
	5-3	DS00387	32C1,2		
49-58	58-49	DS05521	32C1,2		
137-70	70-94	DS06718	32C1,2		
131-33	Not present	DS09345	32C1,2		CC LIGHT + ~50 SECONDARY SITES
90-45	45-89	DS04313	32D1	32D2	
64-62	62-64	DS05920	32D1	32D2	
42-28	28-42	DS02634	32E1	32E2	
63-53	53-63	DS05055	32E1	32E2	
136-55	55-93	DS05277	32E1	32E2	
24-56	56-24	DS05304	32E1	32E2	
	5-21	DS00405	32F1	32F2	
78-2	2-78	DS00174	33A1	33A2	
	5-86	DS00470	33A1	33A2	
24-36	36-24	DS03384	33A1	33A2	
50-72	72-50	DS06866	33A1	33A2	
64-86	86-64	DS08224	33A1	33A2	
45-65	65-45	DS06189	33A3	33A8	
42-46	46-42	DS04362	33A8		
43-26	26-43	DS02443	33B1	33B2	
63-74	74-63	DS07071	33B1	33B2	
	1-39	DS00039	33B3	33C1,2	
	9-21	DS00789	33B12	33C1,2	MAYBE TO 33B11
82-2	2-82	DS00178	33C1,2	33C6	
	4-11	DS00299	33C1,2	33C6	ALSO ~25 WEAK SECONDARY SITES
43-60	60-43	DS05707	33C1,2	33C6	
63-54	54-63	DS05151	33D3	33D4	
63-75	75-63	DS07167	33D3	33D4	
49-79	79-49	DS07537	33D3	33D4	
65-45	45-65	DS04289	33D4	33E1,2	MAYBE TO 33D3
50-85	85-50	DS08114	33E1,2		
65-85	85-65	DS08129	33E4	33E7	MAYBE TO 33E3 &/OR 33E8
90-23	23-89	DS02201	33E9	33E10	
	3-23	DS00215	33F1	33F2	ALSO CC LIGHT
43-20	20-43	DS01867	33F1	33F2	
49-25	25-49	DS02353	34A1	34A2	CC LIGHT
138-36	36-95	DS03455	34A1	34A2	
49-75	75-49	DS07153	34A1	34A2	CC LIGHT
65-83	83-65	DS07937	34A2	34A4	MAYBE TO 34A1
24-9	9-24	DS00792	34A2	34A5	MAYBE TO 34A1 &/OR 34A6
90-83	83-89	DS07961	34A3		
139-6	6-96	DS00576	34A6	34A11	MAYBE TO 34A5
90-60	60-89	DS05753	34B2	34B6	MAYBE 34B1 &/OR 34B7 & CC LITE & ~25 MORE SITES
61-2	2-61	DS00157	34C1	34C2	ALSO CC LIGHT + ~5O LIGHT SECONDARY SITES
	2-5	DS00101	34C4,5		ALSO CC LIGHT
48-84	84-48	DS08016	34C4,5		
24-15	15-24	DS01368	34C4,5	34D1,2	

CAMBR	BERK	DS No.	BEGIN	END	COMMENTS
64-1	1-64	DS00064	34D1,2		
65-64	64-65	DS06113	34D1,2		ALSO CC LIGHT + ~50 SITES
90-86	86-89	DS08249	34D1,2		
135-89	89-92	DS08540	34D1,2		
	9-29	DS00797	34D1,2	34D4	
63-44	44-63	DS04191	34D1,2	34D6	
81-31	Not present	DS09247	34D3	34D8	ALSO ~50 SECONDARY SITES
90-51	51-89	DS04889	34D4	34D6	
64-34	34-64	DS03232	34E1	34E2	
	2-35	DS00131	34E4	34E5	
84-2	2-84	DS00180	34E4	34E5	
44-5	5-44	DS00428	34E4	34E5	
25-44	44-25	DS04153	34E4	34E5	SAME AS DS04249
25-45	45-25	DS04249	34E4	34E5	SAME AS DS04153
25-30	30-25	DS02809	34F1	34F2	
62-40	40-62	DS03806	34F1	34F2	
43-62	62-43	DS05899	34F1	34F2	
65-72	72-65	DS06881	34F1	34F2	SAME AS DS07073
65-74	74-65	DS07073	34F1	34F2	SAME AS DS06881
136-41	41-93	DS03933	34F2	34F3	MAYBE TO 34F1 &/OR 34F4
62-85	85-62	DS08126	34F2	34F3	MAYBE TO 34F1 &/OR 34F4
	8-18	DS00690	34F3	34F4	
25-10	10-25	DS00889	34F3	34F4	
81-63	Not present	DS09279	35A1,2	35A4	
48-40	40-48	DS03792	35A1,2		
135-73	73-92	DS07004	35A3	35A4	
	7-17	DS00593	35A3	35B5	
137-65	65-94	DS06238	35B2	35B3	MAYBE TO 35B1
62-75	75-62	DS07166	35B2	35B6	MAYBE TO 35B1 &/OR 35B7
41-81	81-41	DS07721	35B2	35B10	MAYBE TO 35B1; COVERS MOST OF LARGE PUFF
63-18	18-63	DS01695	35B3	35B6	MAYBE TO 35B7
42-9	9-42	DS00810	35B6	35B7	SAME CLONE AS DS03690(?)
46-16	16-46	DS01486	35B6	35B7	
42-39	39-42	DS03690	35B6	35B7	SAME CLONE AS DS00810(?)
137-91	91-94	DS08734	35B6	35B7	ALSO CC MODERATE + ~50 SECONDARY SITES
	4-20	DS00308	35B8	35B9	
25-88	88-25	DS08377	35B8	35B10	
65-10	10-65	DS00929	35B8	35C1	
50-96	96-50	DS09170	35B9	35C1	MAYBE TO 35B8
46-51	51-46	DS04846	35B10		
62-20	20-62	DS01886	35C1	35C2	MAYBE TO 35C3 + CC MODERATE
24-34	34-24	DS03192	35C1	35C2	MAYBE TO 35C3
138-76	76-95	DS07295	35C1	35C2	
48-93	93-48	DS08880	35C4		
	3-8	DS00200	35C4	35C5	
43-34	34-43	DS03211	35D3	35D4	
25-71	71-25	DS06745	35D3	35D4	ALSO ~50 SECONDARY SITES
62-15	15-62	DS01406	35E1	35E2	
47-32	32-47	DS03023	35E1	35E2	ALSO CC WITH DOTS + ~20 SECONDARY SITES
44-24	24-44	DS02252	35E3	35E5	MAYBE TO 35E6
24-89	89-24	DS08472	35E3	35E5	MAYBE TO 35E6
49-10	10-49	DS00913	35F1,2		
46-24	24-46	DS02254	35F1,2		

CAMBR	BERK	DS No.	BEGIN	END	COMMENTS
1-49	49-1	DS04609	35F1,2		
62-51	51-62	DS04862	35F1,2		
43-55	55-43	DS05227	35F1,2		
137-78	78-94	DS07486	35F1,2		
41-91	91-41	DS08681	35F1,2		
131-75	Not present	DS09387	35F1,2		
	4-77	DS00365	35F1,2	35F4	MAYBE TO 35F5
131-46	Not present	DS09358	35F4	35F5	
	5-52	DS00436	36A1	36A2	
	8-74	DS00746	36A1	36A2	
135-29	29-92	DS02780	36A1	36A2	
136-79	79-93	DS07581	36A1	36A2	
47-80	80-47	DS07631	36A1	36A2	
	2-1	DS00097	36A3		ALSO SOME SECONDARY SITES
46-58	58-46	DS05518	36A6	36A7	
136-66	66-93	DS06333	36A6	36A7	
43-76	76-43	DS07243	36A6	36A7	
	7-16	DS00592	36A7	36A10	MAYBE TO 36A6 &/OR 36A11
51-3	3-51	DS00243	36A10	36B1,2	
	2-39	DS00135	36B1,2		
	7-53	DS00629	36B1,2		
11-37	37-11	DS03467	36B1,2		cDNA 7h1
4-68	68-4	DS06436	36B1,2		cDNA 7h1
25-84	84-25	DS07993	36B1,2		
	2-31	DS00127	36C1	36C2	ALSO CC LIGHT + ~50 SECONDARY SITES
	3-28	DS00220	36C1	36C2	
44-16	16-44	DS01484	36C1	36C2	
136-36	36-93	DS03453	36C1	36C2	
46-90	90-46	DS08590	36C1	36C2	ALSO CC LIGHT + ~25 EUCHROMATIC SITES
44-8	8-44	DS00716	36C3	36C11	
135-42	42-92	DS04028	36C3	36C11	
	7-77	DS00653	36C5	36C11	MAYBE TO 36C4 + CC MODERATE
81-83	Not present	DS09299	36D2	36D3	MAYBE INTO 36D1; DOTS OF WEAK HYBRIDIZATION
	3-80	DS00272	36D3		
	9-93	DS00861	36D3		
50-15	15-50	DS01394	36D3		
45-53	53-45	DS05037	36D3		
50-60	60-50	DS05714	36D3		SAME AS DS06098
50-64	64-50	DS06098	36D3		SAME AS DS05714
43-67	67-43	DS06379	36D3		
47-78	78-47	DS07439	36D3		
137-85	85-94	DS08158	36D3		
65-86	86-65	DS08225	36D3		
138-90	90-95	DS08639	36D3		
74-2	2-74	DS00170	36D3	36E1	
	6-7	DS00487	36E1	36E2	
	6-74	DS00554	36E1	36E2	
81-78	Not present	DS09294	36E1	36E2	
44-6	6-44	DS00524	36E2	36E3	MAYBE TO 36E1 &/OR 36E4
63-34	34-63	DS03231	36E2	36E3	MAYBE TO 36E1 &/OR 36E4
47-82	82-47	DS07823	36E2	36E3	MAYBE TO 36E1 &/OR 36E4
62-33	33-62	DS03134	36F1	36F2	
24-63	63-24	DS05976	36F3	36F9	MAYBE TO 36F10 + CC LIGHT + ~10 WEAK SITES

CAMBR	BERK	DS No.	BEGIN	END	COMMENTS
135-4	4-92	DS00380	36F7	37A1	MAYBE TO 36F6 &/OR 37A2
	8-90	DS00762	37B1,2		
	9-9	DS00777	37B1,2	37B5	
	9-22	DS00790	37B9	37C1,2	MAYBE TO 37B8
65-89	89-65	DS08513	37B9	37C1,2	MAYBE TO 37B8
	8-82	DS00754	37C1,2		
139-36	36-96	DS03456	37D1		
64-70	70-64	DS06688	37D1	37D2	
	1-16	DS00016	37D4	37E1,2	
138-27	27-95	DS02591	37E1,2		
47-55	55-47	DS05231	37E1,2		
81-18	Not present	DS09234	37E1,2		
43-89	89-43	DS08491	37F1	37F6	
92-2	2-91	DS00187	38A1	38A2	ALSO ~15 LIGHT SECONDARY SITES
135-25	25-92	DS02396	38A1	38A2	
24-61	61-24	DS05784	38A1	38A2	ALSO SOME OTHER WEAK SITES
41-95	95-41	DS09065	38A1	38A2	
70-2	2-70	DS00166	38A5	38A6	
45-81	81-45	DS07725	38B2	38B6	MAYBE TO 38B1
	9-95	DS00863	38B3	38B6	
135-10	10-92	DS00956	38C1		
54-3	3-54	DS00246	38C1	38C2	ALSO CC LIGHT + SOME WEAK SECONDARY SITES
	6-26	DS00506	38C1	38C2	
63-41	41-63	DS03903	38C1	38C2	ALSO ~50 SECONDARY SITES
50-44	44-50	DS04178	38C1	38C2	
	6-18	DS00498	38C2	38C5	MAYBE TO 38C1 &/OR 38C6
	6-56	DS00536	38C6	38D1	MAYBE TO 38C5 &/OR 38D2
	3-72	DS00264	38D1	38D2	
48-10	10-48	DS00912	38D1	38D2	
138-55	55-95	DS05279	38D1	38D2	ALSO ~25 SECONDARY EUCHROM SITES
45-60	60-45	DS05709	38D1	38D2	ALSO ~50 SECONDARY SITES
45-74	74-45	DS07053	38D1	38D2	
64-88	88-64	DS08416	38D1	38D2	
139-93	93-96	DS08928	38D1	38D2	
136-22	22-93	DS02109	38D3		
90-48	48-89	DS04601	38E1,2		
	3-3	DS00195	38E1,2	38E5	MAYBE TO 38E6
90-44	44-89	DS04217	38E1,2	38E5	
	7-6	DS00582	38E6	38E9	MAYBE TO 38E5
139-73	73-96	DS07008	38F1	38F3	MAYBE TO 38F4
	3-21	DS00213	38F5	38F6	
135-19	19-92	DS01820	39A3		ALSO SOME VERY WEAK SECONDARY SITES
42-54	54-42	DS05130	39A3	39A7	
131-36	Not present	DS09348	39A3	39B1,2	
	4-71	DS00359	39B1,2		
25-70	70-25	DS06649	39B1,2		ALSO CC LIGHT + FEW SECONDARY SITES OTHER BASES
65-80	80-65	DS07649	39B1,2		
131-42	Not present	DS09354	39B1,2		
43-85	85-43	DS08107	39B3		
	4-13	DS00301	39B3	39C4	
24-17	17-24	DS01560	39C1,2	39D1	MAYBE TO 39D2
81-60	Not present	DS09276	39D1	39D5	
45-72	72-45	DS06861	39D2	39D5	MAYBE TO 39D1

CAMBR	BERK	DS No.	BEGIN	END	COMMENTS
90-27	27-89	DS02585	39D2	39E1	MAYBE TO 39D1 &/OR 39E2
	6-52	DS00532	39D2	39E3,4	MAYBE TO 39D1 + CC LIGHT + ~50 SECONDARY SITES
45-39	39-45	DS03693	39D3	39D5	
47-38	38-47	DS03599	39E2	40A1	MAYBE TO 39E1 &/OR 40A2
49-54	54-49	DS05137	39E5	39F3	
137-59	59-94	DS05662	39E5	39F3	
42-78	78-42	DS07434	40A1	40A4	CC HEAVY
81-13	Not present	DS09229	40A1	40A4	CC ALSO IN SOME NUCLEI & ALL OF CHROMOSOME 4
44-94	94-44	DS08972	40A2	40A3	MAYBE 40A1;40A4 & CC MODERATE & ~50 OTHER SITES
45-96	96-45	DS09165	40A2	40A3	MAYBE TO 40A1 &/OR 40A4
	7-59	DS00635	40A2	40A6	MAYBE TO 40A1
45-66	66-45	DS06285	40B		
	9-5	DS00773	40B2	40B3	MAYBE TO 40B1 &/OR 40B4; ALSO CC + ~100 LITE SITES
	3-26	DS00218	40C1	40C2	
65-32	32-65	DS03041	40C1	40C2	ALSO ~10 WEAK SECONDARY SITES
131-20	Not present	DS09332	40C1	40C2	CC LIGHT + MANY LIGHT SITES
	4-3	DS00291	40D		ALSO CC LIGHT TO MODERATE
	8-84	DS00756	40D		ALSO CC MODERATE
	3-16	DS00208	41C		ALSO CC LIGHT
	6-39	DS00519	41C		ALSO CC LIGHT
90-78	78-89	DS07481	41C		BAND AT BASE OF 2R
63-67	67-63	DS06399	41C1	41C6	CC LIGHT
24-54	54-24	DS05112	41D1,2		CC LIGHT
46-56	56-46	DS05326	41D1,2		ALSO CC LIGHT
137-86	86-94	DS08254	41D1,2		ALSO CC MODERATE
45-22	22-45	DS02061	41D3	41D4	CC LIGHT & SOME WEAK BANDS ON OTHER CHSM BASES
90-76	76-89	DS07289	41E3	41E6	
46-25	25-46	DS02350	41E5		
49-35	35-49	DS03313	41E5	41F1	MAYBE TO 41F2
42-79	79-42	DS07530	41F		CC HEAVY
56-1	1-56	DS00056	41F1	41F2	
	3-75	DS00267	41F1	41F2	ALSO CC LIGHT
	8-85	DS00757	41F1	41F2	ALSO CC MODERATE + ~50 SECONDARY SITES
	8-95	DS00767	41F1	41F2	ALSO CC MODERATE + ~20 LIGHT SITES
50-35	35-50	DS03314	41F1	41F2	
63-40	40-63	DS03807	41F1	41F2	
	7-58	DS00634	41F2	41F3	MAYBE TO 41F1 + CC MODERATE + ~50 SITES
90-79	79-89	DS07577	41F3		
	2-36	DS00132	42A1	42A2	
41-15	15-41	DS01385	42A1	42A2	
137-26	26-94	DS02494	42A1	42A2	
64-39	39-64	DS03712	42A1	42A2	
139-71	71-96	DS06816	42A1	42A2	ALSO CC LITE + ~25 EUCHROM SITES
91-2	2-90	DS00186	42A4,5		ALSO CC LIGHT + MANY LIGHT SITES
50-18	18-50	DS01682	42A8	42A16	
45-56	56-45	DS05325	42A8	42A16	ALSO ~15 SECONDARY SITES
42-73	73-42	DS06954	42A8	42A16	
81-22	Not present	DS09238	42A8	42A16	3-5 WEAK SECONDARY SITES
41-78	78-41	DS07433	42A14	42A17?	THE ? IS SORSA'S DESIGNATION
135-79	79-92	DS07580	42A16	42A18,19	
135-68	68-92	DS06524	42B4	42C2	
131-2	Not present	DS09314	42E5,6	42F1	MAYBE TO 42F2
136-74	74-93	DS07101	42E5,6	42F3	

CAMBR	BERK	DS No.	BEGIN	END	COMMENTS
42-29	29-42	DS02730	43A1		ALSO TIP OF 4 & TIP OF 2R
64-84	84-64	DS08032	43A2	43B1	MAYBE TO 43B2 + ~50 SECONDARY SITES
	2-30	DS00126	43B1	43B2	
45-2	2-45	DS00141	43B1	43B2	
43-92	92-43	DS08779	43B1	43B2	
137-6	6-94	DS00574	43C1,2	43C5,6	
137-36	36-94	DS03454	43D1,2	43D7	
	5-40	DS00424	43D3,4	43E1,2	
47-1	1-47	DS00047	43E1,2	43E4	MAYBE TO 43E5
	7-33	DS00609	43E1,2	43E4	MAYBE TO 43E5
	9-82	DS00850	43E1,2	43E4	MAYBE TO 43E5
47-37	37-47	DS03503	43E5	43E9,10	MAYBE TO 43E4 + 61B1;61B2 IS WEAK SECONDARY SITE
42-59	59-42	DS05610	43E12	43F2	ALSO 5-10 WEAK SECONDARY SITES
136-44	44-93	DS04221	44A1	44A2	
43-78	78-43	DS07435	44A1	44A2	
131-10	Not present	DS09322	44A2	44B1	MAYBE TO 44A1
46-86	86-46	DS08206	44B1		
	7-82	DS00658	44B1	44B2	
131-84	Not present	DS09396	44C1,2	44C4,5	
	2-22	DS00118	44C4,5		
	4-74	DS00362	44C4,5		
	7-91	DS00667	44C4,5		
65-28	28-65	DS02657	44C4,5		
63-57	57-63	DS05439	44C4,5		
45-71	71-45	DS06765	44C4,5		
63-92	92-63	DS08799	44C4,5		
62-96	96-62	DS09182	44C4,5		
48-15	15-48	DS01392	44C4,5	44D1	MAYBE TO 44D2
66-1	1-66	DS00066	44D1	44D2	
44-10	10-44	DS00908	44D1	44D2	
64-64	64-64	DS06112	44D1	44D2	CC LIGHT
139-64	64-96	DS06144	44D1	44D2	
138-69	69-95	DS06623	44D5	44E1,2	
	5-56	DS00440	44F1	44F2	ALSO CC WITH 2-3 DOTS + ~3-4 LIGHT SITES
137-14	14-94	DS01342	44F1	44F2	
47-44	44-47	DS04175	44F1	44F2	ALSO CC LIGHT + ~ 10 SECONDARY SITES
139-45	45-96	DS04320	44F1	44F2	
139-51	51-96	DS04896	44F1	44F2	CC LITE + ~15 SITES; =DS04992, DS05184, DS05280
139-52	52-96	DS04992	44F1	44F2	CC LITE + ~15 SITES; =DS04896, DS05184, DS05280
139-54	54-96	DS05184	44F1	44F2	CC LITE + ~15 SITES; =DS04896, DS04992, DS05280
139-55	55-96	DS05280	44F1	44F2	CC LITE + ~15 SITES; =DS04896, DS04992, DS05184
24-64	64-24	DS06072	44F1	44F2	
41-71	71-41	DS06761	44F9,10		
47-81	81-47	DS07727	45A1		
46-63	63-46	DS05998	45A1	45A2	
	4-12	DS00300	45A7,8	45B1,2	
81-55	Not present	DS09271	45B4	45C1,2	
137-50	50-94	DS04798	45C1,2		
46-57	57-46	DS05422	45C1,2		
	7-81	DS00657	45D1,2		SIGNAL VERY LIGHT
62-64	64-62	DS06110	45D1,2		ALSO CC LIGHT
	9-37	DS00805	45D4	45E1,2	
	7-63	DS00639	45E1,2		ALSO ~50 SECONDARY SITES

CAMBR	BERK	DS No.	BEGIN	END	COMMENTS
135-21	21-92	DS02012	45F1,2	46A1,2	
50-1	1-50	DS00050	46A1,2		
24-7	7-24	DS00600	46A1,2		
138-2	2-95	DS00191	46A1,2	46B1,2	ALSO CC LIGHT + ~15 SECONDARY SITES
47-29	29-47	DS02735	46A1,2	46B1,2	ALSO CC LIGHT + ~15 SECONDARY SITES
135-64	64-92	DS06140	46A1,2	46B1,2	ALSO ~25 WEAK SECONDARY SITES
	7-55	DS00631	46B1,2		ALSO CC LIGHT + SOME LIGHT SECONDARY SITES
47-16	16-47	DS01487	46B1,2		ALSO CC LIGHT + ~10 SECONDARY SITES
90-56	56-89	DS05369	46B1,2		
46-91	91-46	DS08686	46B1,2		ALSO CC LIGHT + ~25 EUCHROMATIC SITES
46-92	92-46	DS08782	46B1,2		
	5-55	DS00439	46B7,8	46C1,2	
90-30	30-89	DS02873	46C1,2		
	6-32	DS00512	46C1,2	46D1,2	
131-61	Not present	DS09373	46C7	46C9	
41-53	53-41	DS05033	46E1,2	46E9?	THE ? IS SORSA'S DESIGNATION
64-49	49-64	DS04672	46E1,2	46F1,2	
90-20	20-89	DS01913	46F5	46F6	
136-54	54-93	DS05181	46F5	46F6	CC LIGHT + ~25 LIGHT SECONDARY SITES
62-68	68-62	DS06494	46F6	46F7	MAYBE TO 46F5
45-87	87-45	DS08301	46F6	46F9	MAYBE TO 46F5 &/OR 46F10 & ~10 OTHER WEAK SITES
	3-30	DS00222	47A1	47A2	
138-82	82-95	DS07871	47A1	47A2	
24-11	11-24	DS00984	47A3	47A7	MAYBE TO 47A8
25-86	86-25	DS08185	47A3	47A7	MAYBE TO 47A8
136-89	89-93	DS08541	47A3	47A7	MAYBE TO 47A8
	8-37	DS00709	47B1,2		
25-18	18-25	DS01657	47B1,2		ALSO 10-12 SECONDARY SITES
49-70	70-49	DS06673	47B1,2	47B4,5	
131-27	Not present	DS09339	47B7,8		
139-41	41-96	DS03936	47C1,2	47C3,4	
	8-10	DS00682	47C1,2	47C7	
68-1	1-68	DS00068	47D1,2		ALSO CC LIGHT + SOME OTHER LIGHT SITES
49-27	27-49	DS02545	47D1,2		
50-27	27-50	DS02546	47D1,2		
81-53	Not present	DS09269	47D1,2	47 D4	ALSO ~100 MODERATELY STRONG SECONDARY SITES
137-62	62-94	DS05950	47D1,2	47D4	
81-36	Not present	DS09252	47D1,2	47D4	
137-44	44-94	DS04222	47D3	47D4	
	7-57	DS00633	47D4		
138-22	22-95	DS02111	47D4		
	8-52	DS00724	47D6	47E1,2	
135-3	3-92	DS00284	47E1,2	47E6	
131-94	Not present	DS09406	47E1,2	47E6	
	3-18	DS00210	47E1,2	47F1,2	
139-24	24-96	DS02304	47F1,2	47F7?	DESIGNATION 47F7? IS SORSA'S; SAME AS DS02784
139-29	29-96	DS02784	47F1,2	47F7?	DESIGNATION 47F7? IS SORSA'S; SAME AS DS02304
86-2	2-86	DS00182	47F1,2	47F8,9	
135-91	91-92	DS08732	47F8,9	47F15,16	MAYBE TO 47F7? (THE ? IS SORSA'S)
135-22	22-92	DS02108	47F8,9	47F18	
137-5	5-94	DS00478	47F15,16	47F18	
68-3	3-68	DS00260	48A1,2	48A6?	48A6? DESIGNATION IS SORSA'S

CAMBR	BERK	DS No.	BEGIN	END	COMMENTS
81-91	Not present	DS09307	48A3,4		ALSO VERY WEAK 52B2;52B3
	6-88	DS00568	48C1	48C2	
	9-56	DS00824	48C1	48C2	
135-56	56-92	DS05372	48C1	48C2	
47-71	71-47	DS06767	48C1	48C2	
	8-35	DS00707	48C5	48D1,2	
	8-91	DS00763	48D1,2		
50-74	74-50	DS07058	48D3	48D8	
63-56	56-63	DS05343	48D5	48E1	MAYBE TO 48E2
44-9	9-44	DS00812	48D6	48E1	MAYBE TO 48E2
139-44	44-96	DS04224	48E8	48F1,2	
139-46	46-96	DS04416	48F1,2	48F5,6	
65-62	62-65	DS05921	48F5,6	49A1	MAYBE TO 49A2
41-1	1-41	DS00041	48F7	49A1	MAYBE TO 49A2
64-81	81-64	DS07744	48F7	49A1	MAYBE TO 49A2
81-1	1-81	DS00081	48F8	49A1	MAYBE TO 49A2
65-95	95-65	DS09089	49A10,11	49B1	MAYBE TO 49B2
137-52	52-94	DS04990	49A10,11	49B1	MAYBE TO 49A10 &/OR 49B2 + CC LITE + ~25 SITES
48-39	39-48	DS03696	49B7	49B12	
	5-71	DS00455	49C1	49D3	ALSO MANY SECONDARY SITES
139-63	63-96	DS06048	49D1	49D2	ALSO CC LITE + ~100 EUCHROM SITES
77-2	2-77	DS00173	49D2	49D3	MAYBE TO 49D1 + MANY WEAK SITES
131-92	Not present	DS09404	49D2	49D3	MAYBE TO 49D1
	5-30	DS00414	49D4?	49E5	DESIGNATION 49D4? IS SORSA'S
24-42	42-24	DS03960	49D4?	49E7	THE ? IS SORSA'S DESIGNATION
62-36	36-62	DS03422	49E6	49F1	MAYBE TO 49F2
64-72	72-64	DS06880	49E6	49F1	MAYBE TO 49F2
65-53	53-65	DS05057	49F1	49F2	
64-60	60-64	DS05728	49F1	49F2	
48-51	51-48	DS04848	50A1,2	50A3,4	
12-56	56-12	DS05292	50A1,2	50A3,4	cDNA 2h10
136-12	12-93	DS01149	50A1,2	50A3,4	MAYBE TO 50A1 &/OR 50A4
48-94	94-48	DS08976	50A1,2	50A3,4	MAYBE TO 50A1 &/OR 50A4
63-88	88-63	DS08415	50A3,4		
	9-70	DS00838	50A5	50A8	
44-2	2-44	DS00140	50A9,10		
	5-67	DS00451	50A9,10	50B1	MAYBE TO 50A9 &/OR 50B2
50-48	48-50	DS04562	50A9,10	50B1	MAYBE TO 50A9 &/OR 50B2
50-62	62-50	DS05906	50B2	50B6,7	MAYBE TO 50B1
43-2	2-43	DS00139	50C1	50C2	
49-77	77-49	DS07345	50C1	50C2	
49-1	1-49	DS00049	50C2	50C3,4	MAYBE TO 50C1
139-1	1-96	DS00096	50C2	50C3,4	MAYBE TO 50C1
137-45	45-94	DS04318	50C2	50C3,4	MAYBE TO 50C1
139-49	49-96	DS04704	50C2	50C3,4	MAYBE TO 50C1
47-52	52-47	DS04943	50C2	50C3,4	MAYBE TO 50C1; =DS05807 & DS05999
47-61	61-47	DS05807	50C2	50C3,4	MAYBE TO 50C1; =DS04943 & DS05999
47-63	63-47	DS05999	50C2	50C3,4	MAYBE TO 50C1; =DS04943 & DS05807
49-91	91-49	DS08689	50C5	50C6,7	
135-31	31-92	DS02972	50C5	50C8	
	7-3	DS00579	50C6,7	50C9,10	
	9-15	DS00783	50C23	50D1,2	
64-10	10-64	DS00928	50D1,2		

CAMBR	BERK	DS No.	BEGIN	END	COMMENTS
50-42	42-50	DS03986	50D1,2		
62-89	89-62	DS08510	50D1,2		
138-1	1-95	DS00095	50E1,2	50E4,5	
44-50	50-44	DS04748	50F1,2	51A1	MAYBE TO 51A2
90-74	74-89	DS07097	50F3	50F9	
	9-89	DS00857	51A6	51A8	
	7-12	DS00588	51B1,2	51B4,5	
25-23	23-25	DS02137	51B1,2	51B4,5	
63-96	96-63	DS09183	51B4,5	51C1	MAYBE TO 51C2
	7-85	DS00661	51C1	51C2	
45-17	17-45	DS01581	51C1	51C2	
62-2	2-62	DS00158	51D1	51D2	
44-52	52-44	DS04940	51D1	51D2	ALSO ~25 OTHER SITES
41-96	96-41	DS09161	51D1	51D2	
131-71	Not present	DS09383	51D7	51E1	MAYBE TO 51D6 &/OR 51E2
136-14	14-93	DS01341	51D8	51D12	ALSO ~15 WEAK SECONDARY SITES
87-1	1-87	DS00087	51E1	51E2	
42-15	15-42	DS01386	51E1	51E2	
44-85	85-44	DS08108	51E3	51E5,6	PROBABLY SAME CLONE AS DS07916
44-83	83-44	DS07916	51E3		PROBABLY SAME CLONE AS DS08108
137-96	96-94	DS09214	51E5,6		
54-2	2-54	DS00150	51E9	51F1,2	
41-52	52-41	DS04937	51E10		
138-42	42-95	DS04031	51F1,2		CC LIGHT
46-84	84-46	DS08014	52A10,11		
137-18	18-94	DS01726	52B1	52B2	MAYBE TO 52B3
50-45	45-50	DS04274	52B1	52B2	MAYBE TO 52B3
136-61	61-93	DS05853	52B1	52B2	MAYBE TO 52B3
137-88	88-94	DS08446	52B1	52B2	MAYBE TO 52B3
48-90	90-48	DS08592	52B2	52B3	
139-30	30-96	DS02880	52B3	52B5	MAYBE TO 52B2
	6-61	DS00541	52C1	52C2	
131-12	Not present	DS09324	52C2	52C6	MAYBE TO 52C1 &/OR 52C7
139-10	10-96	DS00960	52D1	52D2	
24-86	86-24	DS08184	52D2	52D6	MAYBE TO 52D1
	7-20	DS00596	52D4	52D14	
50-84	84-50	DS08018	52D4	52D15	
	7-83	DS00659	52E1,2	52E5,6	
81-26	Not present	DS09242	52F3	52F11	
64-11	11-64	DS01024	53A1		
	6-37	DS00517	53A1	53A2	
131-82	Not present	DS09394	53A2	53B3	MAYBE TO 53A1 &/OR 53B4
	7-73	DS00649	53C1	53C2	
91-1	1-90	DS00090	53C2	53C3	MAYBE TO 53C1 &/OR 53C4
	6-8	DS00488	53C2	53C3	MAYBE TO 53C1 &/OR 53C4
63-32	32-63	DS03039	53C2	53C3,4	MAYBE TO 53C1
45-30	30-45	DS02829	53C4	53C5	MAYBE TO 53C3 &/OR 53C6
131-63	Not present	DS09375	53C5	53C9	MAYBE TO 53C10
137-34	34-94	DS03262	53C7	53C9	MAYBE TO 53C6 &/OR 53C10
135-40	40-92	DS03836	53C7	53C14	MAYBE TO 53C6 &/OR 53C15
135-82	82-92	DS07868	53D1	53D2	
131-67	Not present	DS09379	53D1	53D2	
137-38	38-94	DS03646	53D2	53D6	MAYBE TO 53D1 &/OR 53D7
41-39	39-41	DS03689	53D2	53D8	MAYBE TO 53D1 &/OR 53D9

CAMBR	BERK	DS No.	BEGIN	END	COMMENTS
41-88	88-41	DS08393	53D3	53D5	
64-92	92-64	DS08800	53D7	53D8	MAYBE TO 53D6 &/OR 53D9
41-63	63-41	DS05993	53D12	53E1	MAYBE TO 53E2
25-77	77-25	DS07321	53E1	53E2	
24-90	90-24	DS08568	53E5	53F1	MAYBE TO 53E4 &/OR 53F2
78-1	1-78	DS00078	53E6	53F4	MAYBE TO 53F5
63-60	60-63	DS05727	54A1	54A2	
131-19	Not present	DS09331	54A1	54A2	
62-1	1-62	DS00062	54A2	54B1	MAYBE TO 54A1 &/OR 54B2
	4-73	DS00361	54B1	54B2	
25-12	12-25	DS01081	54B1	54B2	
138-20	20-95	DS01919	54B1	54B2	
49-44	44-49	DS04177	54B1	54B2	
81-43	Not present	DS09259	54B3	54B8	
	7-75	DS00651	54C1,2	54C7,8	
138-94	94-95	DS09023	54C1,2	54C7,8	ALSO CC LIGHT + ~25 SECONDARY SITES
41-37	37-41	DS03497	54C7,8		
136-33	33-93	DS03165	54D1	54D2	
49-78	78-49	DS07441	54D1	54D2	
48-2	2-48	DS00144	54E1	54E7	MAYBE TO 54E8 + WEAK SIGNAL AT 66A8;66A9
43-23	23-43	DS02155	54E8	54F1,2	MAYBE TO 54E7
41-70	70-41	DS06665	54F3	55A1	MAYBE TO 55A2
50-13	13-50	DS01202	55A1	55A2	
	4-33	DS00321	55A2	55A3,4	MAYBE TO 55A1 + SOME VERY LIGHT SITES
81-20	Not present	DS09236	55A2	55B2,3	MAYBE TO 55A1
131-45	Not present	DS09357	55A2	55B2,3	MAYBE TO 55A1
41-51	51-41	DS04841	55A3,4		
	7-1	DS00577	55B1	55B2	
46-13	13-46	DS01198	55B1	55B2,3	
62-54	54-62	DS05150	55B1	55B2,3	
43-25	25-43	DS02347	55B2,3		
44-40	40-44	DS03788	55B2,3		
45-85	85-45	DS08109	55B2,3		
135-18	18-92	DS01724	55C1	55C2	
65-27	27-65	DS02561	55C2	55C4	MAYBE TO 55C1 &/OR 55C5
64-21	21-64	DS01984	55C4	55C5	ALSO ~10 WEAK SITES
	9-53	DS00821	55E1	55E2	ALSO ~6 LIGHT SECONDARY SITES
67-1	1-67	DS00067	55E6,7	55F3	
64-9	9-64	DS00832	55E6,7		
138-95	95-95	DS09119	55E9	55F1	MAYBE TO 55E8 &/OR 55F2
44-86	86-44	DS08204	55F1	55F2	SAME CLONE AS DS08300
44-87	87-44	DS08300	55F1	55F2	SAME CLONE AS DS08204
	3-83	DS00275	55F4	55F13	
42-21	21-42	DS01962	55F6	55F13	
	5-4	DS00388	56A1	56A2	
65-78	78-65	DS07457	56A1	56A2	
137-75	75-94	DS07198	56A2	56B1	MAYBE 56A1 &/OR 56B2 + WK 21D3 + WK 34E4;34E5
65-15	15-65	DS01409	56B1	56B2	
136-34	34-93	DS03261	56B1	56B2	
53-3	3-53	DS00245	56C1	56C2	
	9-86	DS00854	56C1	56C2	
64-36	36-64	DS03424	56C1	56C2	
49-62	62-49	DS05905	56C2	56C4	MAYBE TO 56C1 &/OR 56C5

CAMBR	BERK	DS No.	BEGIN	END	COMMENTS
	5-90	DS00474	56C2	56C6	MAYBE TO 56C1 &/OR 56C7
	5-11	DS00395	56C3	56C6	MAYBE TO 56C7
24-49	49-24	DS04632	56C7	56C11	MAYBE TO 56C6
45-29	29-45	DS02733	56C8	56C11	
	1-20	DS00020	56D3	56D10	
	6-91	DS00571	56D3	56D10	SAME AS DS00572
43-46	46-43	DS04363	56D3	56D10	
	6-40	DS00520	56D10	56E1	MAYBE TO 56E2
137-81	81-94	DS07774	56D11	56E1	MAYBE TO 56E2
63-12	12-63	DS01119	56D11	56E6	
47-59	59-47	DS05615	56D11	56E6	
	5-49	DS00433	56D14	56E1	MAYBE TO 56D13 &/OR 56E2
138-60	60-95	DS05759	56D15	56E1	MAYBE TO 56D12 &/OR 56E2
90-73	73-89	DS07001	56E1	56E2	
90-58	58-89	DS05561	56E3	56E5	
48-75	75-48	DS07152	56F1	56F2	
81-10	Not present	DS09226	56F1	56F9	ALSO ~25 EUCHR SITES
	8-49	DS00721	56F2	56F3	MAYBE TO 56F1 &/OR 56F4
50-39	39-50	DS03698	56F2	56F8	MAYBE TO 56F1 &/OR 56F9
137-72	72-94	DS06910	56F2	56F8	MAYBE TO 56F1 &/OR 56F9 + CC LITE + 25 SITES
	6-4	DS00484	56F4	56F6	MAYBE TO 56F3 &/OR 56F7
	7-28	DS00604	56F4	56F6	MAYBE TO 56F3 &/OR 56F7 + SOME SECONDARY SITES
81-40	Not present	DS09256	56F5	56F9	
	4-81	DS00369	56F7	56F8	MAYBE TO 56F6 &/OR 56F9
43-37	37-43	DS03499	56F8	56F9	
45-67	67-45	DS06381	56F8	56F9	
46-89	89-46	DS08494	56F8	56F9	
50-94	94-50	DS08978	56F8	56F9	
	1-15	DS00015	56F13	56F17	
	3-19	DS00211	57A1	57A2	
	4-4	DS00292	57A1	57A2	
41-62	62-41	DS05897	57A1	57A2	
139-67	67-96	DS06432	57A1	57A2	ALSO CC LITE + ~100 EUCHROM SITES
81-52	Not present	DS09268	57A1	57A2	
25-43	43-25	DS04057	57A3	57A4	
42-26	26-42	DS02442	57A5	57A6	ALSO LITE CC
45-23	23-45	DS02157	57A5	57A7	
138-61	61-95	DS05855	57A5	57A10	
	6-35	DS00515	57A9	57B2	MAYBE TO 57A8 &/OR 57B3 + CC LIGHT
	3-27	DS00219	57B1	57B2	MAYBE TO 57B3
62-26	26-62	DS02462	57B2	57B3	ALSO CC MODERATE + SOME WEAK SITES
50-11	11-50	DS01010	57B5	57B6	
42-69	69-42	DS06570	57B5	57B6	
65-82	82-65	DS07841	57B5	57B6	
81-14	Not present	DS09230	57B5	57B6	
136-25	25-93	DS02397	57B6	57B13	MAYBE TO 57B5 &/OR 57B14
63-47	47-63	DS04479	57B6	57B13	MAYBE TO 57B5 &/OR 57B14
47-41	41-47	DS03887	57B6	57B9,10	MAYBE TO 57B5
137-9	9-94	DS00862	57B13	57B14	
64-25	25-64	DS02368	57B14	57C1	MAYBE TO 57B13 &/OR 57C2
137-37	37-94	DS03550	57B14	57C1	MAYBE TO 57B13 &/OR 57C2
137-56	56-94	DS05374	57B14	57C1	MAYBE TO 57B13 &/OR 57C2
47-91	91-47	DS08687	57C5	57D1	MAYBE TO 57D2

CAMBR	BERK	DS No.	BEGIN	END	COMMENTS
64-2	2-64	DS00160	57D2	57D3	MAYBE TO 57D1 &/OR 57D4
	4-79	DS00367	57D2	57D3	MAYBE TO 57D1 &/OR 57D4
63-69	69-63	DS06591	57D2	57D3	MAYBE TO 57D1 &/OR 57D4
131-13	Not present	DS09325	57D2	57D3	MAYBE TO 57D1 &/OR 57D4
24-51	51-24	DS04824	57D2	57D8,9	MAYBE TO 57D1
	3-35	DS00227	57D11	57D12	
28-83	83-28	DS07900	57D11	57D12	
	9-33	DS00801	57D12	57D13	MAYBE TO 57D11
50-6	6-50	DS00530	57E1	57E2	
136-43	43-93	DS04125	57E1	57E2	
49-60	60-49	DS05713	57E1	57E2	
44-84	84-44	DS08012	57E1	57E2	
	2-13	DS00109	57E3,4	57E9,10	
49-86	86-49	DS08209	57E3,4	57F1	MAYBE TO 57F2
	6-54	DS00534	57F1	57F2	
135-48	48-92	DS04604	57F1	57F2	
64-42	42-64	DS04000	57F4	57F10	
	7-72	DS00648	57F8	58A1	MAYBE TO 58A2 + CC MODERATE + ~250 2NDRY SITES
	8-57	DS00729	58A1	58A2	
138-24	24-95	DS02303	58A1	58A2	ALSO CC MODERATE + ~100 EUCHROM SITES
136-31	31-93	DS02973	58A1	58A2	
41-73	73-41	DS06953	58A1	58A2	
44-4	4-44	DS00332	58A2	58A3	MAYBE TO 58A1 &/OR 58A4
138-63	63-95	DS06047	58A2	58A3	MAYBE TO 58A1 &/OR 58A4
47-18	18-47	DS01679	58A2	58B1	MAYBE TO 58A1 &/OR 58B2
65-70	70-65	DS06689	58A2	58B1	MAYBE TO 58A1 &/OR 58B2
62-79	79-62	DS07550	58A3	58A4	
135-93	93-92	DS08924	58A3	58A4	
	7-66	DS00642	58A4	58B1	MAYBE TO 58A3 &/OR 58B2
46-9	9-46	DS00814	58A4	58B1	MAYBE TO 58A3 &/OR 58B2
49-28	28-49	DS02641	58B1	58B2	
50-28	28-50	DS02642	58B1	58B2	
64-74	74-64	DS07072	58B7	58C1	MAYBE TO 58B6 &/OR 58C2
137-51	51-94	DS04894	58B7	58C5,6	MAYBE TO 58B6
44-29	29-44	DS02732	58C1	58C2	
	7-86	DS00662	58C2	58C7	MAYBE TO 58C1
49-30	30-49	DS02833	58C2	58C7	MAYBE TO 58C1
136-64	64-93	DS06141	58C2	58C7	MAYBE TO 58C1
50-30	30-50	DS02834	58C2	58D1	MAYBE TO 58C1 &/OR 58D2
	6-86	DS00566	58D1	58D2	
50-21	21-50	DS01970	58D1	58D2	
46-41	41-46	DS03886	58D1	58D2	
	9-80	DS00848	58D2	58D4	MAYBE TO 58D1 &/OR 58D5; ALSO 62E9 LIGHT SIGNAL
131-16	Not present	DS09328	58D4	58D5	
136-51	51-93	DS04893	58E1	58E2	
135-9	9-92	DS00860	58E2	58E5	MAYBE TO 58E1
41-59	59-41	DS05609	58F2	58F7	MAYBE TO 58F1
135-23	23-92	DS02204	58F7	59A1	
136-88	88-93	DS08445	59A1		
	4-60	DS00348	59A1	59A2	MAYBE TO 59A3
43-84	84-43	DS08011	59A1	59A2	MAYBE TO 59A3
	3-91	DS00283	59A2	59A3	
49-85	85-49	DS08113	59A2	59A3	

CAMBR	BERK	DS No.	BEGIN	END	COMMENTS
41-60	60-41	DS05705	59A4		
	1-5	DS00005	59B4,5	59B6	MAYBE TO 59B7
55-1	1-55	DS00055	59C2	59C3	MAYBE TO 59C1 &/OR 59C4 + ~50 SECONDARY SITES
136-69	69-93	DS06621	59C5		ALSO ~25 WEAK SECONDARY SITES
	8-63	DS00735	59D1	59D2	
	4-51	DS00339	59D2	59D3	MAYBE TO 59D1 &/OR 59D4
	5-72	DS00456	59D2	59D3	MAYBE TO 59D1 &/OR 59D4
139-12	12-96	DS01152	59D2	59D3	MAYBE TO 59D1 &/OR 59D4
138-39	39-95	DS03743	59D2	59D3	MAYBE TO 59D1 AND/OR 59D4
65-63	63-65	DS06017	59D2	59D3	MAYBE TO 59D1 &/OR 59D4 + CC LIGHT
131-40	Not present	DS09352	59D2	59D3	MAYBE TO 59D1 &/OR 59D4
	8-6	DS00678	59D3	59D4	SAME AS DS00697?
	8-25	DS00697	59D3	59D4	SAME AS DS00678?
	8-58	DS00730	59D3	59D4	
64-32	32-64	DS03040	59D3	59D4	
	5-27	DS00411	59D4	59D6	MAYBE TO 59D3
137-10	10-94	DS00958	59E1	59E2	
42-25	25-42	DS02346	59E1	59E2	ALSO ~25 SECONDARY SITES
24-37	37-24	DS03480	59E1	59E2	
59-3	3-59	DS00251	59F1	59F2	
48-59	59-48	DS05616	59F2	59F4	MAYBE TO 59F1
43-75	75-43	DS07147	59F2	59F4	MAYBE TO 59F1
81-17	Not present	DS09233	59F3	60A1	MAYBE TO 60A2
48-42	42-48	DS03984	60A1	60A2	
42-81	81-42	DS07722	60A1	60A2	
	8-20	DS00692	60A3	60A5	MAYBE TO 60A6
42-64	64-42	DS06090	60A5	60A12	PRIMARY SITE
	7-25	DS00601	60A9	60A10	MAYBE TO 60A8 &/OR 60A11; ~25 OTHER SITES
41-65	65-41	DS06185	60B1	60B2	
	6-63	DS00543	60B2	60B10	MAYBE TO 60B1 &/OR 60B11
136-27	27-93	DS02589	60B6	60B13	
	5-62	DS00446	60C1	60C2	ALSO ~100 SECONDARY SITES
25-17	17-25	DS01561	60C1	60C2	
25-48	48-25	DS04537	60C1	60C2	ALSO ~50 HEAVY SECONDARY SITES
25-69	69-25	DS06553	60C1	60C2	ALSO ~ 50 SECONDARY SITES = DS07033 AND DS07417
25-74	74-25	DS07033	60C1	60C2	ALSO ~ 50 SECONDARY SITES = DS06553 AND DS07417
25-78	78-25	DS07417	60C1	60C2	ALSO ~ 50 SECONDARY SITES = DS06553 AND DS07033
	4-66	DS00354	60C2	60C5	MAYBE TO 60C1
46-54	54-46	DS05134	60C2	60C7	MAYBE TO 60C1 AND/OR 60C8
42-44	44-42	DS04170	60C3	60C4	
137-17	17-94	DS01630	60C7	60C8	
43-53	53-43	DS05035	60D1	60D2	
62-84	84-62	DS08030	60D2	60D5	MAYBE TO 60D1 &/OR 60D6
	1-8	DS00008	60D4	60D7	
48-22	22-48	DS02064	60D5,6	60D10,11	
	8-70	DS00742	60D9	60D10,11	MAYBE TO 60D8
25-4	4-25	DS00313	60D10,11		
43-11	11-43	DS01003	60D10,11		
138-62	62-95	DS05951	60D10,11	60D13	MAYBE TO 60D10 &/OR 60D14 + ~25 MORE SITES
42-52	52-42	DS04938	60E1	60E2	
64-66	66-64	DS06304	60E2	60E4	MAYBE TO 60E1 &/OR 60E5,6
	9-20	DS00788	60E2	60E5,6	MAYBE TO 60E1
137-42	42-94	DS04030	60E3	60E5,6	
64-22	22-64	DS02080	60E5,6	60E8,9	

CAMBR	BERK	DS No.	BEGIN	END	COMMENTS
	3-20	DS00212	60F1	60F5	
62-11	11-62	DS01022	60F1	60F5	
64-16	16-64	DS01504	60F3	60F5	MAYBE TO 60F2
	6-83	DS00563	60F5		
131-89	Not present	DS09401	60F5		
47-34	34-47	DS03215	61A1	61A2	VERY TIP OF 3L ONLY
81-59	Not present	DS09275	61A1	61A3	
44-31	31-44	DS02924	61A3	61A6	
24-60	60-24	DS05688	61A5	61B1	MAYBE TO 61B2
	7-7	DS00583	61B1	61B2	
24-88	88-24	DS08376	61B2	61B3	MAYBE TO 61B1
131-5	Not present	DS09317	61B3	61C1,2	
63-31	31-63	DS02943	61C1,2		
64-93	93-64	DS08896	61C1,2		
	6-59	DS00539	61C1,2	61C3	
	7-47	DS00623	61C1,2	61C3	
62-39	39-62	DS03710	61C1,2	61C4	
46-11	11-46	DS01006	61C3	61C4	
	8-89	DS00761	61C6	61C8	MAYBE TO 61C5
	8-19	DS00691	61C7	61C9	
	8-21	DS00693	61C7	61C9	
131-39	Not present	DS09351	61C7	61C9	
135-96	96-92	DS09212	61C8		
131-65	Not present	DS09377	61C8		
	2-7	DS00103	61D1	61D2	
	5-85	DS00469	61D1	61D2	
	7-68	DS00644	61D1	61D2	
48-44	44-48	DS04176	61D1	61D2	
131-70	Not present	DS09382	61D1	61D2	
42-13	13-42	DS01194	61E1		
47-48	48-47	DS04559	61E1		ALSO CC LIGHT + ~150 SECONDARY SITES
48-26	26-48	DS02448	61E1	61E2	
64-82	82-64	DS07840	61E2	61F2	ALSO ~50 LIGHT SITES
44-39	39-44	DS03692	61F1		
43-81	81-43	DS07723	61F1	61F4	
50-73	73-50	DS06962	61F3	61F4	
55-2	2-55	DS00151	62A1	62A2	ALSO CC LIGHT + ~25 SECONDARY SITES
46-29	29-46	DS02734	62A1	62A2	
44-36	36-44	DS03404	62A1	62A2	
41-40	40-41	DS03785	62A2	62A4	MAYBE TO 62A1
41-23	23-41	DS02153	62A3	62A8	
41-55	55-41	DS05225	62A4	62A9	MAYBE TO 62A10
42-17	17-42	DS01578	62A9	62A10	
81-42	Not present	DS09258	62A9	62A10	
90-29	29-89	DS02777	62A10	62B1,2	MAYBE TO 62A9
	7-37	DS00613	62B1,2		
43-39	39-43	DS03691	62B1,2	62B3	
	5-41	DS00425	62B1,2	62B7,8	
41-25	25-41	DS02345	62B3	62B4,5	SAME AS DS02825
41-30	30-41	DS02825	62B3	62B4,5	SAME AS DS02345
46-53	53-46	DS05038	62B4,5	62B6	
90-71	71-89	DS06809	62B7,8	62B10	MAYBE TO 62B11
51-2	2-51	DS00147	62B10	62B11	
45-3	3-45	DS00237	62B10	62B11	

CAMBR	BERK	DS No.	BEGIN	END	COMMENTS
	9-43	DS00811	62B10	62B11	
28-21	21-28	DS01948	62B10	62B11	
64-35	35-64	DS03328	62B10	62B11	
43-88	88-43	DS08395	62B10	62B11	
	6-62	DS00542	62B12	62C3	
32-75	75-32	DS07136	62C1		CC MOD HVY & ~25 OTHER SITES; cDNA 3h12
46-44	44-46	DS04174	62C1	62C2	
139-77	77-96	DS07392	62C1	62C2	
75-2	2-75	DS00171	62C1	62C3	
	8-30	DS00702	62C1	62C3	
3-16	16-3	DS01443	62C1	62C3	cDNA 3h12
42-42	42-42	DS03978	62C1	62C3	
44-71	71-44	DS06764	62C4		
47-23	23-47	DS02159	62C4	62D1	MAYBE TO 62D2 + CC LABELED WITH WEAK DOTS
43-13	13-43	DS01195	62D1	62D2	
45-34	34-45	DS03213	62D1	62D2	
48-89	89-48	DS08496	62D1	62D2	
48-18	18-48	DS01680	62D2	62D4	MAYBE TO 62D1
49-18	18-49	DS01681	62D5	62D6	
131-91	Not present	DS09403	62E5		
	5-54	DS00438	62E5	62E7	
64-50	50-64	DS04768	62E5	62E7	
131-3	Not present	DS09315	62E5	62E8	
139-28	28-96	DS02688	62E8	62E9	
24-78	78-24	DS07416	62E9	62F1,2	
	6-43	DS00523	62F1,2		ALSO CC LIGHT
	8-68	DS00740	62F1,2		
49-29	29-49	DS02737	62F1,2		
50-29	29-50	DS02738	62F1,2		
	9-6	DS00774	62F1,2	62F6	ALSO CC LIGHT + SOME SECONDARY SITES
90-26	26-89	DS02489	62F3	62F6	
42-83	83-42	DS07914	62F4	63A1	MAYBE TO 63A2
135-33	33-92	DS03164	62F6	63A1	
90-2	2-89	DS00185	63A1		
42-53	53-42	DS05034	63A1	63A2	
	4-58	DS00346	63A2	63A3	MAYBE TO 63A1
135-30	30-92	DS02876	63A3	63A5	
135-12	12-92	DS01148	63A3	63A7	
62-34	34-62	DS03230	63A3	63B1,2	
35-17	17-35	DS01571	63A6	63A8	cDNA 9a12
25-83	83-25	DS07897	63A6	63B7	MAYBE TO 63B8
35-96	96-35	DS09155	63B1,2	63B8	cDNA 9a12
85-1	1-85	DS00085	63D1	63D2	
138-29	29-95	DS02783	63D1	63D2	
43-80	80-43	DS07627	63D1	63D3	
49-12	12-49	DS01105	63D2	63D3	MAYBE TO 63D1
138-87	87-95	DS08351	63D2	63D3	MAYBE TO 63D1
	4-19	DS00307	63D3		
79-1	1-79	DS00079	63E1		
49-42	42-49	DS03985	63E1		
47-31	31-47	DS02927	63E1	63E4	MAYBE TO 63E5
45-80	80-45	DS07629	63E1	63E4	MAYBE TO 63E5
	6-28	DS00508	63E4	63E5	

CAMBR	BERK	DS No.	BEGIN	END	COMMENTS
	9-72	DS00840	63E4	63E5	
42-65	65-42	DS06186	63E4	63E5	
90-1	1-89	DS00089	63E6	63E9	
81-32	Not present	DS09248	63E6	63E9	
42-11	11-42	DS01002	63F1,2	63F7	
	8-56	DS00728	63F6	64A1	MAYBE TO 64A2
136-49	49-93	DS04701	64A1		
44-51	51-44	DS04844	64A1	64A2	
50-52	52-50	DS04946	64A1	64A2	
	5-5	DS00389	64A4	64A5	
	8-39	DS00711	64A4	64A5	
64-63	63-64	DS06016	64A5	64A12	MAYBE TO 64A4
65-17	17-65	DS01601	64B4	64B11	
24-87	87-24	DS08280	64B6	64B9,10	MAYBE TO 64B5 + ~30 OTHER WEAK SITES
81-57	Not present	DS09273	64B6	64B9,10	
131-25	Not present	DS09337	64B6	64B9,10	MAYBE TO 64B5
137-28	28-94	DS02686	64B9,10	64B12	MAYBE TO 64B13
139-11	11-96	DS01056	64B12	64B13	
	8-69	DS00741	64B13	64B16	MAYBE TO 64B12
48-38	38-48	DS03600	64B13	64C1	MAYBE TO 64B12 &/OR 64C2
43-36	36-43	DS03403	64B16	64C1	MAYBE TO 64C2
	1-25	DS00025	64C1	64C2	
	8-3	DS00675	64C1	64C2	
44-56	56-44	DS05324	64C1	64C2	
44-78	78-44	DS07436	64C1	64C2	SAME CLONE AS DS07628
44-80	80-44	DS07628	64C1	64C2	SAME CLONE AS DS07436
24-55	55-24	DS05208	64C2	64C4	MAYBE TO 64C1 &/OR 64C5
44-79	79-44	DS07532	64C3	64C4	MAYBE TO 64C5
	4-2	DS00290	64C4	64C5	
25-22	22-25	DS02041	64C4	64C5	
44-45	45-44	DS04268	64C4	64C5	
46-1	1-46	DS00046	64C6	64C8	
42-94	94-42	DS08970	64C6	64C8	
24-72	72-24	DS06840	64C9,10	64C11	MAYBE TO 64C12
62-10	10-62	DS00926	64D1	64D2	
42-12	12-42	DS01098	64D1	64D2	
45-31	31-45	DS02925	64D1	64D2	
62-72	72-62	DS06878	64D1	64D2	
	5-51	DS00435	64E1	64E2	
43-64	64-43	DS06091	64E1	64E2	
137-73	73-94	DS07006	64E1	64E2	
136-78	78-93	DS07485	64E11,12		
65-20	20-65	DS01889	64F1	64F2	
43-28	28-43	DS02635	64F1	64F2	
	2-14	DS00110	64F2	64F3	MAYBE TO 64F1
49-93	93-49	DS08881	65A6	65A10	MAYBE TO 65A5 &/OR 65A11
	4-7	DS00295	65A11	65B1	MAYBE TO 65A10 &/OR 65B2
	6-79	DS00559	65B1	65B2	
135-44	44-92	DS04220	65B1	65B2	
	4-18	DS00306	65B3	65B5	
49-87	87-49	DS08305	65C1,2		
	3-17	DS00209	65C3	65D3	
24-26	26-24	DS02424	65C4	65D1	MAYBE TO 65D2

CAMBR	BERK	DS No.	BEGIN	END	COMMENTS
	2-15	DS00111	65D2	65D3	MAYBE TO 65D1
	4-86	DS00374	65D2	65D3	MAYBE TO 65D1
81-88	Not present	DS09304	65D2	65D3	MAYBE INTO 65D1 & CC LIGHT-MODERATE & ~25 SITES
138-37	37-95	DS03551	65D2	65D6	MAYBE TO 65D1
50-65	65-50	DS06194	65D4	65E1	MAYBE TO 65E2
47-40	40-47	DS03791	65E1	65E2	
63-17	17-63	DS01599	65E2	65E3	MAYBE TO 65E1 &/OR 65E4
45-69	69-45	DS06573	65E3	65E4	
138-84	84-95	DS08063	65E4	65E7	MAYBE TO 65E3
48-3	3-48	DS00240	65E8	65E9	ALSO WEAK SIGNAL AT 66A8;66A9
	5-68	DS00452	65E8	65E9	
	6-17	DS00497	65E8	65E9	
41-22	22-41	DS02057	65E8	65E9	
138-49	49-95	DS04703	65E8	65E9	
49-57	57-49	DS05425	65E8	65E9	
70-1	1-70	DS00070	65E9	65E12	MAYBE TO 65E8
24-12	12-24	DS01080	65F1	65F2	
49-65	65-49	DS06193	65F1	65F2	
131-38	Not present	DS09350	65F1	65F2	
43-79	79-43	DS07531	65F6	66A1	MAYBE TO 66A2
139-17	17-96	DS01632	66A1	66A2	SAME AS DS01728
139-18	18-96	DS01728	66A1	66A2	SAME AS DS01632
48-53	53-48	DS05040	66A1	66A2	
	9-49	DS00817	66A5	66A6	
49-21	21-49	DS01969	66A5	66A6	
45-61	61-45	DS05805	66A5	66A6	
63-64	64-63	DS06111	66A5	66A6	
49-66	66-49	DS06289	66A5	66A6	
62-58	58-62	DS05534	66A6	66A8	MAYBE TO 66A5 &/OR 66A9
138-93	93-95	DS08927	66A6	66A8	MAYBE TO 66A5 &/OR 66A9
47-94	94-47	DS08975	66A6	66A8	MAYBE TO 66A5 &/OR 66A9
	5-23	DS00407	66A8	66A9	
45-54	54-45	DS05133	66A8	66A9	
47-96	96-47	DS09167	66A8	66A9	ALSO WEAK SIGNAL AT 61E1
49-33	33-49	DS03121	66A12	66A14	MAYBE TO 66A11
	9-36	DS00804	66A12	66A17	MAYBE TO 66A11 &/OR 66A18 + ~150 HEAVY SITES
50-33	33-50	DS03122	66A12	66A17	MAYBE TO 66A11 &/OR 66A18
	3-87	DS00279	66A14	66A17	MAYBE TO 66A18
48-76	76-48	DS07248	66A15	66A21	MAYBE TO 66A14 + ~20 LIGHT SECONDARY SITES
43-66	66-43	DS06283	66A17	66A18	
50-16	16-50	DS01490	66A19	66A20	
	7-45	DS00621	66B1	66B2	
46-78	78-46	DS07438	66B2	66B5	MAYBE TO 66B1 AND/OR 66B6
63-2	2-63	DS00159	66B6	66B10	MAYBE TO 66B5 &/OR 66B11
48-9	9-48	DS00816	66C1,2		ALSO WEAK SIGNAL AT 74A1;74A2
63-85	85-63	DS08127	66C1,2		
65-93	93-65	DS08897	66C1,2		ALSO CC VERY LIGHT + ~5-10 VERY WEAK SITES
136-39	39-93	DS03741	66C3	66C7	
	7-90	DS00666	66C5	66C8,9	
	6-30	DS00510	66C8,9	66D1,2	
	7-8	DS00584	66C8,9	66D1,2	ALSO SOME WEAK SECONDARY SITES
	7-30	DS00606	66C8,9	66D1,2	
	9-30	DS00798	66C8,9	66D1,2	

CAMBR	BERK	DS No.	BEGIN	END	COMMENTS
48-83	83-48	DS07920	66D1,2	66D4	
137-4	4-94	DS00382	66D5	66D6	
	5-8	DS00392	66D5	66D6	
	6-49	DS00529	66D5	66D6	
49-36	36-49	DS03409	66D5	66D6	
50-36	36-50	DS03410	66D5	66D6	
65-42	42-65	DS04001	66D5	66D6	
135-52	52-92	DS04988	66D10	66D11	
	2-27	DS00123	66D11	66D12	MAYBE TO 66D10
	9-28	DS00796	66D11	66D12	MAYBE TO 66D10
	9-7	DS00775	66D11	66D15	MAYBE TO 66D10
	6-57	DS00537	66E1	66E2	
25-31	31-25	DS02905	66E1	66E2	
137-60	60-94	DS05758	66E1	66E2	
65-94	94-65	DS08993	66E4	66F1	MAYBE TO 66F2
	1-35	DS00035	66F1	66F2	
	2-19	DS00115	66F1	66F2	
	5-33	DS00417	66F1	66F2	
4-79	79-4	DS07492	66F4	67A1,2	cDNA 2g8
63-70	70-63	DS06687	67A1,2		
48-82	82-48	DS07824	67A1,2		
24-82	82-24	DS07800	67A1,2	67A3	MAYBE TO 67A4 + ~200 SECONDARY SITES
30-11	11-30	DS00990	67A1,2	67A4	cDNA 2g8
65-29	29-65	DS02753	67A3	67A4	
24-75	75-24	DS07128	67A3	67A4	
	9-8	DS00776	67A5	67B4	MAYBE TO 67B5
13-43	43-13	DS04045	67A5	67B5	CC MOD HVY & ~30 OTHER SITES; cDNA 3f12
11-50	50-11	DS04715	67A5	67B5	CC MOD HVY & ~30 OTHER SITES; cDNA 3f12
90-41	41-89	DS03929	67B2	67B7	MAYBE TO 67B1 &/OR 67B8
139-61	61-96	DS05856	67B2	67B7	MAYBE TO 67B1 &/OR 67B8 + CC LITE + ~25 SITES
	4-52	DS00340	67B6	67B11	
44-34	34-44	DS03212	67B7	67B8	
	4-15	DS00303	67B8	67B11	MAYBE TO 67B7
	2-17	DS00113	67C1	67C2	ALSO ~50 SECONDARY SITES
64-3	3-64	DS00256	67C1	67C2	ALSO CC LIGHT + SOME LIGHT SITES
	5-28	DS00412	67C1	67C2	
	8-13	DS00685	67C1	67C2	ALSO SOME SECONDARY SITES
65-13	13-65	DS01217	67C1	67C2	ALSO ~ 50 SECONDARY SITES; POSSIBLY =DS01697
65-18	18-65	DS01697	67C1	67C2	ALSO ~ 10 WEAK SITES; POSSIBLY SAME AS DS01217
90-53	53-89	DS05081	67C2	67C4	MAYBE TO 67C1
137-95	95-94	DS09118	67C4		
41-93	93-41	DS08873	67C4	67C8	
60-1	1-60	DS00060	67C10		
	8-11	DS00683	67C10		
	1-10	DS00010	67D1	67D2	
	9-65	DS00833	67D1	67D2	
139-74	74-96	DS07104	67D1	67D2	
25-68	68-25	DS06457	67D3	67D5,6	
69-1	1-69	DS00069	67D11	67D12	
	5-78	DS00462	67D11	67D12	
	4-14	DS00302	67D12	67E1	MAYBE TO 67D11 &/OR 67E2
	7-34	DS00610	67D12	67E1	MAYBE TO 67D11 &/OR 57E2
	1-30	DS00030	67D13		

CAMBR	BERK	DS No.	BEGIN	END	COMMENTS
65-14	14-65	DS01313	67D13		
48-78	78-48	DS07440	67D13		
50-19	19-50	DS01778	67D13	67E1	MAYBE TO 67E2
	5-26	DS00410	67E1	67E2	ALSO CC VERY LIGHT
	5-83	DS00467	67E1	67E2	
81-7	Not present	DS09223	67E1	67E4	
90-34	34-89	DS03257	67E3	67E4	
136-87	87-93	DS08349	67E3	67E4	
24-2	2-24	DS00120	67F1,2		
	4-22	DS00310	67F1,2		
62-46	46-62	DS04382	67F1,2	67F3	
18-49	49-18	DS04626	67F1,2	67F3	cDNA 8h7
25-52	52-25	DS04921	67F1,2	67F3	
65-19	19-65	DS01793	68A1	68A2	
11-24	24-11	DS02219	68A1	68A2	cDNA 8h7
45-49	49-45	DS04653	68A1	68A2	
81-81	Not present	DS09297	68A1	68A2	
50-40	40-50	DS03794	68A2	68A6	MAYBE TO 68A1
48-68	68-48	DS06480	68A3	68A4	MAYBE TO 68A5
	8-59	DS00731	68A3	68A7	
	6-12	DS00492	68A5	68B1,2	MAYBE TO 68A4
86-1	1-86	DS00086	68B1,2		
	8-15	DS00687	68C2	68C3	MAYBE TO 68C1 &/OR 68C4
47-47	47-47	DS04463	68C2	68C3	MAYBE TO 68C1 &/OR 68C4
136-10	10-93	DS00957	68C10	68C12	MAYBE TO 68C13 + ~10 WEAK SECONDARY SITES
135-85	85-92	DS08156	68C10	68C12	MAYBE TO 68C13
137-15	15-94	DS01438	68C12	68C13	
	6-33	DS00513	68D1	68D2	
43-51	51-43	DS04843	68D3		
	1-31	DS00031	68D4,5		ALSO CC LIGHT + SOME LIGHT SECONDARY SITES
	2-28	DS00124	68D4,5		
52-2	2-52	DS00148	68D4,5		ALSO CC LIGHT + ~50 LIGHT SECONDARY SITES
	4-37	DS00325	68D4,5		FAINT SIGNAL AND ONLY IN SOME NUCLEI
24-91	91-24	DS08664	68D4,5		CC LIGHT + ~20 WEAK SECONDARY SITES
136-57	57-93	DS05469	68E1	68E2	
73-2	2-73	DS00169	68E2	68E3	MAYBE TO 68E1
131-69	Not present	DS09381	68E2	68E3	MAYBE TO 68E1
65-36	36-65	DS03425	68F1	68F2	
48-62	62-48	DS05904	68F1	68F2	
75-1	1-75	DS00075	68F+3		BAND DESIGNATION IS SORSA'S
81-74	Not present	DS09290	68F+3	68F6	BAND DESIGNATION FROM SORSA
131-6	Not present	DS09318	68F+3	68F8	68F+3 TERMINOLOGY IS SORSA'S
	2-3	DS00099	69C2	69C11	MAYBE TO 69C1
64-29	29-64	DS02752	69C2	69C8	MAYBE TO 69C1
	3-93	DS00285	69C3	69C11	
138-78	78-95	DS07487	69C3	69C11	
64-89	89-64	DS08512	69D1		
44-1	1-44	DS00044	69D1	69D2	MAYBE TO 69D3
137-84	84-94	DS08062	69D1	69D2	MAYBE TO 69D3
	4-10	DS00298	69D2	69D3	
73-1	1-73	DS00073	69D4	69E5	
	9-2	DS00770	69D4	69E6,7	
46-64	64-46	DS06094	69E6,7		

CAMBR	BERK	DS No.	BEGIN	END	COMMENTS
63-77	77-63	DS07359	69F2	69F7	
63-94	94-63	DS08991	69F2	69F7	MAYBE TO 69F1
63-45	45-63	DS04287	69F2	70A1	MAYBE TO 69F1 &/OR 70A2
42-30	30-42	DS02826	69F3	69F7	
46-4	4-46	DS00334	70A1	70A2	
42-74	74-42	DS07050	70A1	70A5	
48-46	46-48	DS04368	70A2	70A4	MAYBE TO 70A1 &/OR 70A5
24-68	68-24	DS06456	70A2	70A4	MAYBE TO 70A1 &/OR 70A5
42-50	50-42	DS04746	70A4	70A5	
90-63	63-89	DS06041	70A4	70A5	
45-37	37-45	DS03501	70B1	70B3	
	3-11	DS00203	70C1	70C2	
24-16	16-24	DS01464	70C1	70C2	CC LIGHT + ~15 VERY WEAK EUCHROMATIC SITES
46-19	19-46	DS01774	70C1	70C2	
25-27	27-25	DS02521	70C1	70C2	
44-46	46-44	DS04364	70C1	70C2	
65-61	61-65	DS05825	70C1	70C2	
64-73	73-64	DS06976	70C1	70C2	
131-62	Not present	DS09374	70C1	70C2	
63-13	13-63	DS01215	70C4	70C6	
137-33	33-94	DS03166	70C4	70C9	
139-34	34-96	DS03264	70C12,13		
64-15	15-64	DS01408	70D1		
44-82	82-44	DS07820	70D1		
	9-83	DS00851	70D1	70D2	
25-81	81-25	DS07705	70D1	70D2	
131-14	Not present	DS09326	70D1	70D2	
44-13	13-44	DS01196	70D3		
64-61	61-64	DS05824	70D4	70D5	ALSO CC LIGHT + SOME WEAK SITES
48-52	52-48	DS04944	70E1	70E2	
139-2	2-96	DS00192	70E4	70E5	
47-88	88-47	DS08399	70E4	70E5	ALSO LIGHT SIGNAL AT 79E1
63-90	90-63	DS08607	70E4	70E5	
138-38	38-95	DS03647	70F1,2		
	7-70	DS00646	70F1,2	70F7	ALSO CC MODERATE + ~25 SECONDARY SITES
65-41	41-65	DS03905	70F3	70F7	ALSO CC LIGHT + SOME SECONDARY SITES
61-1	1-61	DS00061	71A1	71A2	
138-81	81-95	DS07775	71A1	71A2	
42-2	2-42	DS00138	71A2	71B1	MAYBE TO 71A1
62-95	95-62	DS09086	71A3		
	6-19	DS00499	71B1	71B2,+	DESIGNATION 71B2,+ IS SORSA'S
50-3	3-50	DS00242	71B3	71B8	
	8-2	DS00674	71B3	71B8	
136-96	96-93	DS09213	71B3	71B8	
81-84	Not present	DS09300	71B3	71C1	MAYBE INTO 71C2
44-19	19-44	DS01772	71B4	71C1	MAYBE TO 71C2
50-37	37-50	DS03506	71B4	71C1	MAYBE TO 71C2
136-48	48-93	DS04605	71B6	71C1	MAYBE TO 71C2
	8-40	DS00712	71C1	71C2	
49-56	56-49	DS05329	71C1	71C2	
42-60	60-42	DS05706	71C1	71C2	CC MODERATELY HEAVY
62-65	65-62	DS06206	71C1	71C2	
81-25	Not present	DS09241	71C1	71C2	

CAMBR	BERK	DS No.	BEGIN	END	COMMENTS
46-85	85-46	DS08110	71C3	71D1	MAYBE TO 71D2
45-15	15-45	DS01389	71D1	71D2	
24-48	48-24	DS04536	71D2	71E1	MAYBE TO 71D1 &/OR 71E2
25-53	53-25	DS05017	71D2	71E1	MAYBE TO 71D1 &/OR 71E2
63-20	20-63	DS01887	71E3	71E5	
135-5	5-92	DS00476	71F1	71F2	
63-22	22-63	DS02079	71F1	71F2	
46-61	61-46	DS05806	71F1	71F2	
	6-53	DS00533	72A1	72A2	
49-50	50-49	DS04753	72A1	72A2	
135-81	81-92	DS07772	72A1	72A2	PROBABLY SAME CLONE AS DS08348
135-87	87-92	DS08348	72A1	72A2	PROBABLY SAME CLONE AS DS07772
46-38	38-46	DS03598	72C1	72D6	
131-58	Not present	DS09370	72D1,2	72D8	
45-36	36-45	DS03405	72D10,11	72D12	
47-28	28-47	DS02639	72D10,11	72E1	MAYBE TO 72E2
131-77	Not present	DS09389	72E1	72E2	ALSO 2-3 SECONDARY SITES
136-30	30-93	DS02877	72E3	72F4	
135-11	11-92	DS01052	72F1,2		
136-37	37-93	DS03549	72F1,2		
25-59	59-25	DS05593	73A1	73A2	
63-14	14-63	DS01311	73A2	73A3	MAYBE TO 73A1 &/OR 73A4,+
135-54	54-92	DS05180	73A2	73A3	MAYBE TO 73A1 &/OR 73A4,+
76-2	2-76	DS00172	73A2	73A9	MAYBE TO 73A1 &/OR 73A10
50-87	87-50	DS08306	73B5,6		
50-46	46-50	DS04370	73B5,6	73C1	MAYBE TO 73C2
3-21	21-3	DS01923	73B5,6	73D1,2	cDNA 3c12
29-82	82-29	DS07805	73B5,6	73D1,2	cDNA 3c12
8-93	93-8	DS08840	73B5,6	73D1,2	cDNA 3c12
	6-9	DS00489	73D1,2	73D6	
	4-95	DS00383	73D4	73E1,2	ALSO CC LIGHT + FEW LIGHT SITES
63-1	1-63	DS00063	73E1,2	73F1	
56-3	3-56	DS00248	74A1	74A2	
62-90	90-62	DS08606	74A1	74A2	
81-15	Not present	DS09231	74A1	74A5	
48-8	8-48	DS00720	74A2	74A4	MAYBE 74A1;74A5 + WEAK SIGNALS AT 66C1,2
	7-49	DS00625	74A4	74A5	
	7-80	DS00656	74A4	74A5	
52-1	1-52	DS00052	74A5	74B1	MAYBE TO 74A4 &/OR 74B2
	4-54	DS00342	74B2	74C1,2	MAYBE TO 74B1
139-69	69-96	DS06624	74C1,2	74D1	MAYBE TO 74D2 AND ALSO ~25 EUCHROM SITES
25-94	94-25	DS08953	74C1,2	74D1	MAYBE TO 74D2
135-59	59-92	DS05660	74D4	74E4	
138-35	35-95	DS03359	74E2	74F1	MAYBE TO 74E1
138-15	15-95	DS01439	75A1	75A2	
65-33	33-65	DS03137	75A1	75A2	ALSO SOME OTHER VERY WEAK SITES
24-45	45-24	DS04248	75A1	75A2	
48-69	69-48	DS06576	75A1	75A2	
49-55	55-49	DS05233	75A3	75B1	MAYBE TO 75B2
	9-26	DS00794	75A4,5	75B1	MAYBE TO 75B2
64-17	17-64	DS01600	75B2	75B5	MAYBE TO 75B1 + ~5-10 WEAK SITES
81-21	Not present	DS09237	75B3	75B11?	THE ? IS SORSA'S DESIGNATION
42-34	34-42	DS03210	75B3	75B7	

CAMBR	BERK	DS No.	BEGIN	END	COMMENTS
136-13	13-93	DS01245	75B3	75B8	
45-52	52-45	DS04941	75B13		
136-21	21-93	DS02013	75C1	75C2	
25-46	46-25	DS04345	75C1	75C2	
138-47	47-95	DS04511	75C1	75C2	
41-58	58-41	DS05513	75C1	75C2	
62-60	60-62	DS05726	75C1	75C2	
65-90	90-65	DS08609	75C1	75C2	
43-91	91-43	DS08683	75C1	75C2	
	6-73	DS00553	75C2	75C4	MAYBE TO 75C1 + CC LIGHT + ~25 SECONDARY SITES
64-69	69-64	DS06592	75C2	75C4	MAYBE TO 75C1; ALSO CC MODERATE
	3-40	DS00232	75C3	75C6	ALSO CC LIGHT + 150-200 SECONDARY SITES
81-45	Not present	DS09261	75C6		
81-58	Not present	DS09274	75C6		
62-83	83-62	DS07934	75D2	75E1	MAYBE TO 75D1 &/OR 75E2
48-73	73-48	DS06960	75D3	75D8	
45-14	14-45	DS01293	75E2	75E7	MAYBE TO 75E1
136-23	23-93	DS02205	75E2	75E7	MAYBE TO 75E1
48-72	72-48	DS06864	75F6	75F7	
24-18	18-24	DS01656	76A2	76A3,4	MAYBE TO 76A1
44-76	76-44	DS07244	76A2	76A3,4	MAYBE TO 76A1
24-79	79-24	DS07512	76A2	76A3,4	MAYBE TO 76A1
43-82	82-43	DS07819	76A2	76A3,4	MAYBE TO 76A1
	7-54	DS00630	76B2	76B3,4	MAYBE TO 76B1
	7-87	DS00663	76B2	76B3,4	MAYBE TO 76B1
25-40	40-25	DS03769	76B2	76B3,4	MAYBE TO 76B1
63-48	48-63	DS04575	76B2	76B3,4	MAYBE TO 76B1 + CC LIGHT + ~10 LIGHT SITES
137-80	80-94	DS07678	76B2	76B3,4	MAYBE TO 76B1
24-93	93-24	DS08856	76B2	76B3,4	MAYBE TO 76B1
	1-4	DS00004	76B3,4		PR0BABLY SAME AS DS00003
44-15	15-44	DS01388	76B3,4		
44-30	30-44	DS02828	76B3,4		
41-86	86-41	DS08201	76B3,4		
81-90	Not present	DS09306	76B5	76B11	
57-2	2-57	DS00153	76B5	76B9,10	
	1-3	DS00003	76B5		PR0BABLY SAME AS DS00004
	6-75	DS00555	76B6	76B11	
	8-14	DS00686	76B6	76B11	
25-92	92-25	DS08761	76B6	76B11	
	1-27	DS00027	76C3	76C6	ALSO CC LIGHT + ~25 LIGHT SECONDARY SITES
42-86	86-42	DS08202	76D1	76D4	
137-66	66-94	DS06334	76D1	76E1	MAYBE TO 76E2
	2-8	DS00104	76F1	76F2	
71-2	2-71	DS00167	76F2	76F3	MAYBE TO 76F1 + FEW WEAK SITES
	4-40	DS00328	77A1	77A2	
	7-14	DS00590	77A1	77A2	
136-32	32-93	DS03069	77A3	77B2	MAYBE TO 77B3
139-3	3-96	DS00288	77A4	77B1,+	DESIGNATION 77B1,+ IS SORSA'S
62-12	12-62	DS01118	77B1,+	77B6,7	77B1,+ IS SORSA'S DESIGNATION; ALSO CC LIGHT
50-69	69-50	DS06578	77B3	77B5	MAYBE TO 77B2; ALSO CC LIGHT
138-59	59-95	DS05663	77B4,+	77B9	77B4,+ IS SORSA'S DESIGNATION FOR THIS BAND
49-15	15-49	DS01393	77B4,+	77C1	CC LIGHT & FEW OTHER SITES

CAMBR	BERK	DS No.	BEGIN	END	COMMENTS
	4-65	DS00353	77C1	77C3	MAYBE TO 77C4
	2-37	DS00133	77E1	77E2	
	7-69	DS00645	77E2	77E3	MAYBE TO 77E1 &/OR 77E4
25-15	15-25	DS01369	77E2	77E3	MAYBE TO 77E1 &/OR 77E4
47-77	77-47	DS07343	77E2	77E3	MAYBE TO 77E1 &/OR 77E4; =DS07535
47-79	79-47	DS07535	77E2	77E3	MAYBE TO 77E1 &/OR 77E4; =DS07343
46-8	8-46	DS00718	77E3	77E4	
47-3	3-47	DS00239	77E4	77F1	MAYBE TO 77E3 &/OR 77F2
44-62	62-44	DS05900	77F3	78A+,1	
	7-84	DS00660	78A+,1	78A2,+	BAND DESIGNATIONS SORSA'S
131-26	Not present	DS09338	78A+,1	78A2,+	BAND DESIGNATIONS ARE SORSA'S
41-83	83-41	DS07913	78A2,+	78A4	MAYBE TO 78A+,1
88-1	1-88	DS00088	78B1	78B2	ALSO CC LIGHT + ~10-15 SECONDARY SITES
65-3	3-65	DS00257	78B1	78B2	
	9-10	DS00778	78B1	78B2	
46-26	26-46	DS02446	78B1	78B2	
131-79	Not present	DS09391	78B1	78B2	ALSO ~5 SECONDARY SITES
46-10	10-46	DS00910	78B2	78C1	MAYBE TO 78B1 &/OR 78C2
	4-90	DS00378	78C4	78C8	
44-3	3-44	DS00236	78D1	78D2	
41-46	46-41	DS04361	78D1	78D2	
139-56	56-96	DS05376	78D1	78D2	
	7-38	DS00614	78D4	78D5	
42-71	71-42	DS06762	78D4	78D5	CC LIGHT
	8-29	DS00701	79A3	79A4	
41-21	21-41	DS01961	79A3	79A4	
43-18	18-43	DS01675	79B1	79B3	
48-95	95-48	DS09072	79B1	79B3	ALSO WEAK SIGNAL AT 99E1;99E2
62-8	8-62	DS00734	79B1	79C1	MAYBE TO 79C2
64-52	52-64	DS04960	79C2	79D1	NAYBE TO 79C1 &/OR 79D2 + 5-6 SECONDARY SITES
64-27	27-64	DS02560	79D1	79D2	
45-42	42-45	DS03981	79D1	79D2	
65-51	51-65	DS04865	79D1	79D2	
49-53	53-49	DS05041	79D1	79D2	
90-61	61-89	DS05849	79D1	79D2	
138-67	67-95	DS06431	79D1	79D2	
131-35	Not present	DS09347	79D1	79D2	
139-21	21-96	DS02016	79D2	79D4	MAYBE TO 79D1
44-38	38-44	DS03596	79D4		
	6-36	DS00516	79E1	79E2	
135-15	15-92	DS01436	79E1	79E2	
47-87	87-47	DS08303	79E1	79E2	
	4-75	DS00363	79E2	79E3	MAYBE TO 79E1 &/OR 79E4
	7-27	DS00603	79E2	79E3	MAYBE TO 79E1 &/OR 79E4 + CC LIGHT + ~50 SITES
64-79	79-64	DS07552	79E2	79E3	MAYBE TO 79E1 &/OR 79E4
131-24	Not present	DS09336	79E2	79E3	MAYBE TO 79E1 &/OR 79E4 + CC LITE + ~25 WK SITES
49-19	19-49	DS01777	79E3	79E4	ALSO ~15 SECONDARY SITES
136-42	42-93	DS04029	79E3	79E4	CC VERY LIGHT + ~10 WEAK SCNDRY SITES (=DS04317)
136-45	45-93	DS04317	79E3	79E4	CC VERY LIGHT + ~10 WEAK SCNDRY SITES (=DS04029)
62-92	92-62	DS08798	79E3	79E4	ALSO ~25 SECONDARY SITES
81-27	Not present	DS09243	79E3	79E4	
90-52	52-89	DS04985	79E4	79E6	MAYBE TO 79E7
	4-43	DS00331	79F1	79F2	

CAMBR	BERK	DS No.	BEGIN	END	COMMENTS
	6-15	DS00495	80A1	80A2	ALSO CC LIGHT
42-36	36-42	DS03402	80A1	80A2	
24-59	59-24	DS05592	80A1	80A2	
139-35	35-96	DS03360	80A3	80A4	ALSO CC LITE; SAME AS DS03552
139-37	37-96	DS03552	80A3	80A4	ALSO CC LITE; SAME AS DS03360
	5-37	DS00421	80B1	80B2	ALSO CC LIGHT + ~5 SECONDARY SITES
138-14	14-95	DS01343	80B1	80B2	ALSO CC LITE TO MODERATE
65-44	44-65	DS04193	80B1	80B2	
	6-25	DS00505	82A1,2		ALSO CC LIGHT + ~25 LIGHT EUCHROMATIC SITES
50-31	31-50	DS02930	82A1,2		
131-7	Not present	DS09319	82A1,2		CC LIGHT + ~20 SECONDARY SITES
135-32	32-92	DS03068	82A3	82A6	ALSO SEVERAL VERY LIGHT SECONDARY SITES
	6-65	DS00545	82A4	82B3	
	7-4	DS00580	82B1		
	5-69	DS00453	82B1	82B4	
	5-81	DS00465	82B1	82B4	
49-39	39-49	DS03697	82B1	82B4	
43-54	54-43	DS05131	82B1	82B4	
136-72	72-93	DS06909	82B1	82B4	
64-71	71-64	DS06784	82B2	82C1	MAYBE TO 82C2
24-44	44-24	DS04152	82C2	82D1	MAYBE TO 82C1
24-77	77-24	DS07320	82C3	82D1	
131-47	Not present	DS09359	82C3	82D2	
48-81	81-48	DS07728	82D5	82D8	MAYBE TO 82D4
57-3	3-57	DS00249	82D7	82E1	MAYBE TO 82E2
	5-1	DS00385	82E1	82E2	
136-62	62-93	DS05949	82E1	82E2	
59-2	2-59	DS00155	82E2	82E4	MAYBE TO 82E1
45-28	28-45	DS02637	82E6	82E7	
	3-36	DS00228	82F1,2		
90-67	67-89	DS06425	82F1,2		
65-21	21-65	DS01985	82F1,2	82F6	
	8-36	DS00708	82F8	82F9	ALSO CC LIGHT + SOME SECONDARY SITES
49-68	68-49	DS06481	83B1	83B6	MAYBE TO 83B7
138-80	80-95	DS07679	83B3	83B6	MAYBE TO 83B7
44-27	27-44	DS02540	83B8	83B9	
81-68	Not present	DS09284	83C3	83C8	
	5-91	DS00475	83C4	83C8	ALSO CC LIGHT
81-79	Not present	DS09295	83C4	83C8	ALSO LIGHT CC & ~10 WEAK SITES
42-57	57-42	DS05418	83C6	83D1	MAY EXTEND TO 83D2; SAME CLONE AS DS05226?
	5-47	DS00431	83D1	83D2	
42-70	70-42	DS06666	83D1	83D5	
25-19	19-25	DS01753	83D3		
65-60	60-65	DS05729	83D3		
	5-17	DS00401	83D4	83D5	
136-95	95-93	DS09117	83E1	83E2	
136-47	47-93	DS04509	83E3	83E4	
41-18	18-41	DS01673	83E4		
	3-71	DS00263	83F1,2		ALSO ~25 LIGHT SECONDARY SITES
45-78	78-45	DS07437	83F1,2		
	2-33	DS00129	84A1	84A2	
47-53	53-47	DS05039	84A1	84A2	
136-70	70-93	DS06717	84A1	84A2	

CAMBR	BERK	DS No.	BEGIN	END	COMMENTS
41-92	92-41	DS08777	84A1	84A2	
	3-84	DS00276	84A4	84A5,+	84A5,+ DESIGNATION IS SORSA'S
138-86	86-95	DS08255	84A4	84A5,+	BAND DESIG SORSA'S + CC MODERATE + ~50 MORE SITES
63-50	50-63	DS04767	84A5,+	84A6	84A5,+ DESIGNATION IS SORSA'S
136-2	2-93	DS00189	84A6		
	5-13	DS00397	84B1	84B2	
	5-80	DS00464	84B2	84C1,2	MAYBE TO 84B1
42-84	84-42	DS08010	84C1,2	84C6	
	7-79	DS00655	84C1,2	84C8	
88-2	2-88	DS00184	84D1	84D2	
90-42	42-89	DS04025	84D1	84D2	
64-45	45-64	DS04288	84D1	84D2	
139-59	59-96	DS05664	84D1	84D2	
90-69	69-89	DS06617	84D1	84D2	
137-94	94-94	DS09022	84D2	84D3,4	MAYBE TO 84D1
49-63	63-49	DS06001	84D3,4		
41-72	72-41	DS06857	84D3,4		
62-88	88-62	DS08414	84D3,4		
43-70	70-43	DS06667	84D5?	84D8	THE ? IS SORSA'S DESIGNATION
45-8	8-45	DS00717	84D6,7		
42-35	35-42	DS03306	84D6,7		
25-65	65-25	DS06169	84D6,7		
135-61	61-92	DS05852	84D6,7	84D8	
62-81	81-62	DS07742	84D6,7	84D9	MAYBE TO 84D10
	3-15	DS00207	84D9	84D10	
	6-58	DS00538	84D9	84D10	
50-7	7-50	DS00626	84D9	84D10	
	8-61	DS00733	84D9	84D10	
65-35	35-65	DS03329	84D9	84D10	
24-66	66-24	DS06264	84D9	84D10	CC MODERATE + ~12-15 EUCHROMATIC SITES
81-46	Not present	DS09262	84D9	84D10	
25-35	35-25	DS03289	84D10	84D11	MAYBE TO 84D9 &/OR 84D12 + CC LIGHT + ~25 SITES
49-49	49-49	DS04657	84D11	84D12	
50-57	57-50	DS05426	84D12	84D14	MAYBE TO 84D11
25-56	56-25	DS05305	84D13	84E1,2	
64-85	85-64	DS08128	84E1,2		
135-62	62-92	DS05948	84E1,2	84E6	
135-27	27-92	DS02588	84E1,2	84E9	
138-11	11-95	DS01055	84E5	84E9	
65-73	73-65	DS06977	84E10,11		
	1-6	DS00006	84F1	84F2	
	9-51	DS00819	84F1	84F2	
44-17	17-44	DS01580	84F1	84F2	
48-19	19-48	DS01776	84F1	84F2	
63-49	49-63	DS04671	84F1	84F2	
45-64	64-45	DS06093	84F1	84F2	
136-26	26-93	DS02493	84F3	84F6,7	
50-81	81-50	DS07730	84F4	84F6,7	
131-55	Not present	DS09367	84F4	84F6,7	
135-74	74-92	DS07100	85A1		
45-46	46-45	DS04365	85A1	85A2	MAYBE TO 85A3
131-59	Not present	DS09371	85A1	85A2	MAYBE TO 85A3 + ~25 WEAK SECONDARY SITES

CAMBR	BERK	DS No.	BEGIN	END	COMMENTS
24-24	24-24	DS02232	85A1	85A5	MAYBE TO 85A6
49-64	64-49	DS06097	85A1	85A5	MAYBE TO 85A6
50-80	80-50	DS07634	85A2	85A3	ALSO CC MODERATE + ~2-3 OTHER SITES
46-96	96-46	DS09166	85A2	85A3	
139-48	48-96	DS04608	85A4		
	6-90	DS00570	85A7	85B1	ALSO CC LIGHT + MANY SECONDARY SITES
	8-26	DS00698	85B1	85B4	MAYBE TO 85B5
43-14	14-43	DS01291	85B1	85B4	MAYBE TO 85B5
90-33	33-89	DS03161	85B1	85B4	MAYBE TO 85B5
90-24	24-89	DS02297	85B4	85B5	
	6-20	DS00500	85C5	85C9	MAYBE TO 85C4 &/OR 85C10
	1-7	DS00007	85C6	85C13	
69-3	3-69	DS00261	85C12	85D1	
65-69	69-65	DS06593	85D1	85D2	
135-69	69-92	DS06620	85D1	85D2	
	2-9	DS00105	85D7,8		
24-53	53-24	DS05016	85D11,12	85D17	
25-62	62-25	DS05881	85D16	85D17	MAYBE TO 85D15
44-33	33-44	DS03116	85D18	85D22	
41-19	19-41	DS01769	85D18	85D27	
47-30	30-47	DS02831	85D21	85D24	MAYBE TO 85D25
25-39	39-25	DS03673	85E2	85E4	MAYBE TO 85E1 &/OR 85E5
49-71	71-49	DS06769	85E9	85E10	
24-8	8-24	DS00696	85E14	85F1	MAYBE TO 85E13 &/OR 85F2
138-56	56-95	DS05375	85F1	85F2	
46-62	62-46	DS05902	85F2	85F4	MAYBE TO 85F1 AND/OR 85F5
	1-13	DS00013	85F5	85F7	MAYBE TO 85F4 &/OR 85F8
49-61	61-49	DS05809	85F8	85F11	MAYBE TO 85F7
65-57	57-65	DS05441	85F8	85F12	MAYBE TO 85F7 &/OR 85F13
137-20	20-94	DS01918	85F13	86A1	MAYBE TO 85F12 &/OR 86A2
136-46	46-93	DS04413	86A1	86A2	
25-64	64-25	DS06073	86A1	86A2	ALSO SOME WEAK SECONDARY SITES
135-50	50-92	DS04796	86A2	86A4	MAYBE TO 86A1 &/OR 86A5 + ~6-10 WK 2NDRY SITES
43-83	83-43	DS07915	86A2	86A4	MAYBE TO 86A1 &/OR 86A5
41-3	3-41	DS00233	86A4	86A5	
	3-81	DS00273	86A4	86A5	
62-28	28-62	DS02654	86A4	86A5	
43-56	56-43	DS05323	86A4	86A5	
49-73	73-49	DS06961	86A5	86B1,2	MAYBE TO 86A4
24-39	39-24	DS03672	86A6	86B6	
	8-86	DS00758	86B1,2	86B6	
74-1	1-74	DS00074	86C1	86C2	
	5-12	DS00396	86C1	86C2	
42-7	7-42	DS00618	86C1	86C2	
	8-16	DS00688	86C1	86C2	
42-19	19-42	DS01770	86C1	86C2	SAME CLONE AS DS00618(?)
43-22	22-43	DS02059	86C1	86C2	
137-55	55-94	DS05278	86C1	86C2	
136-59	59-93	DS05661	86C1		
	8-87	DS00759	86C7,8		ALSO CC LIGHT + SOME LIGHT SECONDARY SITES
42-14	14-42	DS01290	86C7,8		PERHAPS ALSO ~25 EUCHR SITES
138-48	48-95	DS04607	86C13	86C15	
42-3	3-42	DS00234	86D1	86D2	ALSO ~25 EUCHROMATIC SITES

CAMBR	BERK	DS No.	BEGIN	END	COMMENTS
	4-38	DS00326	86D1	86D2	
	6-55	DS00535	86D1	86D2	
	9-58	DS00826	86D1	86D2	
46-45	45-46	DS04270	86D1	86D2	
47-64	64-47	DS06095	86D1	86D2	
44-88	88-44	DS08396	86D1	86D2	
46-88	88-46	DS08398	86D1	86D2	
49-92	92-49	DS08785	86D1	86D2	
42-93	93-42	DS08874	86D1	86D2	
45-50	50-45	DS04749	86D1	86FD2	
63-80	80-63	DS07647	86D3	86D10	
43-77	77-43	DS07339	86D5	86D10	MAYBE TO 86D4
25-28	28-25	DS02617	86E1	86E2	CC LIGHT + ~200 SECONDARY SITES
46-39	39-46	DS03694	86E1	86E2	SAME AS DS03790
46-40	40-46	DS03790	86E1	86E2	SAME AS DS03694
64-40	40-64	DS03808	86E1	86E2	
43-93	93-43	DS08875	86E1	86E2	
	6-80	DS00560	86E7	86E8	
	6-95	DS00575	86E7	86E8	
139-43	43-96	DS04128	86E7	86E8	
48-55	55-48	DS05232	86E8	86E13	MAYBE TO 86E7; SAME AS DS05424
48-57	57-48	DS05424	86E8	86E13	MAYBE TO 86E7; SAME AS DS05232
81-16	Not present	DS09232	86E8	86E9	
131-74	Not present	DS09386	86E9	86E16	
63-15	15-63	DS01407	86E10	86E14	
24-4	4-24	DS00312	86E13	86E17	MAYBE TO 86E12
	3-13	DS00205	86F1,2		
	8-75	DS00747	86F1,2		
	4-72	DS00360	86F1,2	86F4	
	5-29	DS00413	86F1,2	86F8	MAYBE TO 86F9
	7-21	DS00597	86F3	86F10	
65-56	56-65	DS05345	86F7	86F8	MAYBE TO 86F6 &/OR 86F9
64-5	5-64	DS00448	87A1	87A2	
62-48	48-62	DS04574	87A1	87A2	
138-25	25-95	DS02399	87A2	87A4	MAYBE TO 87A1 &/OR 87A5
64-12	12-64	DS01120	87A4	87A5	
90-80	80-89	DS07673	87A4	87A5	SAME AS DS07769
90-81	81-89	DS07769	87A4	87A5	SAME AS DS07673
63-46	46-63	DS04383	87B1	87B2	
45-55	55-45	DS05229	87B1	87B2	
	2-18	DS00114	87B2	87B4	MAYBE TO 87B1 &/OR 87B5
50-79	79-50	DS07538	87B2	87B4	MAYBE TO 87B1 &/OR 87B5
25-49	49-25	DS04633	87B4	87B5	
48-66	66-48	DS06288	87B4	87B5	
25-91	91-25	DS08665	87B4	87B5	
138-17	17-95	DS01631	87B5	87B8	MAYBE TO 87B4 AND/OR 87B9
24-74	74-24	DS07032	87B5	87B8	MAYBE TO 87B4 &/OR 87B9
24-13	13-24	DS01176	87B8	87B9	
	4-31	DS00319	87B9	87B15	MAYBE TO 87B8
62-52	52-62	DS04958	87B9	87C1	MAYBE TO 87B8
136-29	29-93	DS02781	87C1		CC LIGHT
24-94	94-24	DS08952	87C1		CC LIGHT
	7-26	DS00602	87C1	87C2	MAYBE TO 87C3

CAMBR	BERK	DS No.	BEGIN	END	COMMENTS
44-75	75-44	DS07148	87C1	87C2	MAYBE TO 87C3
	3-22	DS00214	87C2	87C3	
	8-9	DS00681	87C2	87C3	
	6-93	DS00573	87C4	87C9	
48-87	87-48	DS08304	87D1	87D2	
	5-74	DS00458	87D2	87D3	MAYBE TO 87D1 &/OR 87D4
48-49	49-48	DS04656	87D2	87D3	MAYBE TO 87D1 &/OR 87D4
138-88	88-95	DS08447	87D2	87D3	MAYBE TO 87D1 &/OR 87D4 + CC LITE + ~25 SITES
46-15	15-46	DS01390	87D3	87D4	
80-2	2-80	DS00176	87E1	87E2	
	5-20	DS00404	87E1	87E2	
139-42	42-96	DS04032	87E1	87E2	
44-59	59-44	DS05612	87E1	87E2	
131-51	Not present	DS09363	87E1	87E2	ALSO 3-4 WEAK SECONDARY SITES
	4-76	DS00364	87E2	87E6	MAYBE TO 87E1
139-23	23-96	DS02208	87E4	87E12	
	8-31	DS00703	87E5	87E6	ALSO CC LIGHT
	9-31	DS00799	87E5	87E6	ALSO CC LIGHT
	3-14	DS00206	87F1	87F2	
64-51	51-64	DS04864	87F1	87F2	
49-41	41-49	DS03889	87F3		
138-44	44-95	DS04223	87F3	87F11	
47-68	68-47	DS06479	87F8	87F12	MAYBE TO 87F7 &/OR 87F13
137-23	23-94	DS02206	87F12	87F13	
138-54	54-95	DS05183	87F12	87F13	
	3-37	DS00229	87F14	88A1	MAYBE TO 88A2 + MANY WEAK SECONDARY SITES
136-40	40-93	DS03837	88A1	88A2	
64-77	77-64	DS07360	88A1	88A2	
47-13	13-47	DS01199	88A2	88A4,5	MAYBE TO 88A1
43-72	72-43	DS06859	88A9	88A12	
46-82	82-46	DS07822	88A10	88A12	ALSO CC LIGHT
	9-35	DS00803	88B1,2		ALSO CC LIGHT
41-89	89-41	DS08489	88B1,2		ALSO CC LIGHT + MANY WEAK SITES
46-94	94-46	DS08974	88B5	88B9	
81-95	Not present	DS09311	88C1	88C2	
136-28	28-93	DS02685	88C6		
	7-74	DS00650	88C9	88C10	
	9-32	DS00800	88C9	88D2	
	9-47	DS00815	88D1	88D2	
90-18	18-89	DS01721	88D1	88D2	
46-47	47-46	DS04462	88D5	88D6	
	5-35	DS00419	88D5	88D9	
	9-16	DS00784	88D5	88D9	
131-56	Not present	DS09368	88D5	88D9	SAME AS DSO9369
131-57	Not present	DS09369	88D5	88D9	SAME AS DSO9368
43-74	74-43	DS07051	88D6	88D9	
	3-78	DS00270	88D8		
25-11	11-25	DS00985	88D8		
65-54	54-65	DS05153	88D8		
65-58	58-65	DS05537	88D8		
49-17	17-49	DS01585	88E1	88E2	ALSO LIGHT SIGNAL AT 26C3
	3-5	DS00197	88E3	88E4	ALSO CC MODERATE + ~30 SECONDARY SITES
63-72	72-63	DS06879	88E3	88E4	

CAMBR	BERK	DS No.	BEGIN	END	COMMENTS
44-18	18-44	DS01676	88E9	88F1,2	
	6-16	DS00496	89A1		
	3-31	DS00223	89A1	89A2	ALSO CC LIGHT + SOME SECONDARY SITES
58-3	3-58	DS00250	89A1	89A2	ALSO ~20 WEAK SECONDARY SITES
46-43	43-46	DS04078	89A1	89A2	
135-43	43-92	DS04124	89A1	89A2	ALSO ~10 WEAK SECONDARY SITES
4-60	60-4	DS05668	89A1	89A2	
48-88	88-48	DS08400	89A1	89A2	
131-22	Not present	DS09334	89A1	89A2	CC LIGHT + ~50 LIGHT SECONDARY SITES
49-31	31-49	DS02929	89A1	89A4	
46-35	35-46	DS03310	89A3		
	9-84	DS00852	89A6	89A8,9	
64-46	46-64	DS04384	89A8,9		
24-50	50-24	DS04728	89A8,9		
62-35	35-62	DS03326	89B2	89B3	
	8-5	DS00677	89B6	89B7	
135-47	47-92	DS04508	89B6	89B7	
41-84	84-41	DS08009	89B9,10	89B12	MAYBE INTO 89B13
139-38	38-96	DS03648	89B11		
135-67	67-92	DS06428	89B13	89B16	MAYBE TO 89B12
24-76	76-24	DS07224	89B13	89B16	MAYBE TO 89B12 &/OR 89B17
50-83	83-50	DS07922	89B19	89C1	MAYBE TO 89B18 &/OR 89C2
62-13	13-62	DS01214	89C1	89C2	
47-39	39-47	DS03695	89C2	89C4,5	MAYBE TO 89C1 + WEAK SIGNAL AT 39E2;40A1
50-63	63-50	DS06002	89C3	89C6	MAYBE TO 89C7
44-26	26-44	DS02444	89C7		
81-49	Not present	DS09265	89D1	89D3	
34-36	36-34	DS03394	89D1	89D9	cDNA 3d9
2-32	32-2	DS02978	89D1	89E4	cDNA 1d10
46-23	23-46	DS02158	89D2	89D3	MAYBE TO 89D1
63-83	83-63	DS07935	89D2	89D4	MAYBE TO 89D1
139-65	65-96	DS06240	89D2	89D6,7	MAYBE TO 89D1
6-59	59-6	DS05574	89D4	89D9	cDNA 1d10
45-86	86-45	DS08205	89D4	89D9	
	5-87	DS00471	89E1	89E2	
	9-52	DS00820	89E1	89E2	
135-13	13-92	DS01244	89E2	89E3	MAYBE TO 89E1 &/OR 89E4
48-36	36-48	DS03408	89E2	89E3	MAYBE TO 89E1 &/OR 89E4
	2-32	DS00128	89E3	89E4	
	6-31	DS00511	89E4	89E7	MAYBE TO 89E3 &/OR 89E8
	9-78	DS00846	89E5	89E6	
47-43	43-47	DS04079	89E8	89E10	MAYBE TO 89E7 &/OR 89E11
25-54	54-25	DS05113	89E8	89E10	MAYBE TO 89E7 &/OR 89E11
43-50	50-43	DS04747	89E10	89E11	
50-20	20-50	DS01874	89F3	90A1	MAYBE TO 90A2
24-20	20-24	DS01848	90A1	90A2	
48-24	24-48	DS02256	90A1	90A2	
62-24	24-62	DS02270	90A1	90A2	
44-72	72-44	DS06860	90A1	90A2	
81-9	Not present	DS09225	90A1	90A2	
25-63	63-25	DS05977	90A2	90B1,2	
44-25	25-44	DS02348	90A3	90A7	
42-37	37-42	DS03498	90A3	90A7	

CAMBR	BERK	DS No.	BEGIN	END	COMMENTS
44-65	65-44	DS06188	90A3	90A7	
	4-17	DS00305	90B1,2		
	6-14	DS00494	90B1,2		
45-33	33-45	DS03117	90C7,8		
	4-35	DS00323	90D1,2		
44-20	20-44	DS01868	90D1,2		
24-6	6-24	DS00504	90D1,2		
136-63	63-93	DS06045	90D1,2		
42-95	95-42	DS09066	90D1,2		
40-25	25-40	DS02344	90D1,2	90E2	cDNA 7f6
47-60	60-47	DS05711	90E1	90E2	
63-89	89-63	DS08511	90E1	90E2	
62-70	70-62	DS06686	90E3	90F1,2	
137-74	74-94	DS07102	90E3	90F4	
48-6	6-48	DS00528	91A1	91A2	
138-74	74-95	DS07103	91A1	91A2	
49-89	89-49	DS08497	91A1	91A2	
45-91	91-45	DS08685	91A3		
139-25	25-96	DS02400	91A3	91A8	
24-32	32-24	DS03000	91B1		
25-20	20-25	DS01849	91B1	91B2	MAYBE TO 91B3
41-14	14-41	DS01289	91C1,2		
41-87	87-41	DS08297	91C1,2		
45-83	83-45	DS07917	91C1,2	91C7	
	4-69	DS00357	91D1	91D2	
63-30	30-63	DS02847	91D1	91D2	
41-50	50-41	DS04745	91D1	91D2	
135-65	65-92	DS06236	91D1	91D2	
62-66	66-62	DS06302	91D1	91D2	
64-68	68-64	DS06496	91D2	91D5	MAYBE TO 91D1
20-21	21-20	DS01940	91E1	91E2	cDNA 1d10
81-44	Not present	DS09260	91E1	91E2	
131-76	Not present	DS09388	91E2	91E4	MAYBE TO 91E1
	4-70	DS00358	91F1	91F2	
131-17	Not present	DS09329	91F1	91F2	
64-59	59-64	DS05632	91F4,5		
90-25	25-89	DS02393	91F10	91F11	
	4-23	DS00311	91F11	91F13	MAYBE TO 91F10 + CC LIGHT + SOME LIGHT SITES
50-51	51-50	DS04850	91F11	91F13	MAYBE TO 91F10
63-23	23-63	DS02175	92A1	92A2	
25-36	36-25	DS03385	92A1	92A2	CC MODERATE + ~200 HEAVY SECONDARY SITES
48-65	65-48	DS06192	92A2	92A3	MAYBE TO 92A1
25-13	13-25	DS01177	92A3	92A5	
47-83	83-47	DS07919	92A7	92A10	MAYBE TO 92A6 &/OR 92A11
137-79	79-94	DS07582	92A11	92B1	MAYBE TO 92A10 &/OR 92B2
90-6	6-89	DS00569	92B1		
50-54	54-50	DS05138	92B1	92B2	ALSO SOME LIGHT SECONDARY SITES
90-68	68-89	DS06521	92B1	92B2	
138-68	68-95	DS06527	92B1	92B2	
62-73	73-62	DS06974	92B2	92B3	MAYBE TO 92B1 + CC LIGHT + A FEW LIGHT SITES
50-43	43-50	DS04082	92B2	92B5	MAYBE TO 92B1
45-21	21-45	DS01965	92B4	92B11	
	7-23	DS00599	92B6,7	92C1,2	

CAMBR	BERK	DS No.	BEGIN	END	COMMENTS
	6-45	DS00525	92D1	92D2	
50-9	9-50	DS00818	92D1	92D2	SAME AS DS01106 AND DS02162
50-12	12-50	DS01106	92D1	92D2	SAME AS DS00818 AND DS02162
50-23	23-50	DS02162	92D1	92D2	SAME AS DS00818 AND DS01106
7-56	56-7	DS05287	92D1	92D2	ALSO ~10 SECONDARY SITES
65-66	66-65	DS06305	92D1	92D2	ALSO CC LIGHT + SOME WEAK SITES
135-77	77-92	DS07388	92D1	92D2	
63-84	84-63	DS08031	92D1	92D2	
41-85	85-41	DS08105	92D1	92D2	
131-43	Not present	DS09355	92D1	92D2	CC MODERATE + ~250 SECONDARY SITES
136-82	82-93	DS07869	92D3	92D9	
50-90	90-50	DS08594	92D3	92D9	
48-20	20-48	DS01872	92D3	92E1	MAYBE TO 92E2
	5-15	DS00399	92E1	92E2	
	9-67	DS00835	92E1	92E2	ALSO 5-6 SECONDARY SITES, ESPECIALLY 19F1;19F2
50-56	56-50	DS05330	92E1	92E2	
44-73	73-44	DS06956	92E1	92E2	SAME CLONE AS DS07052
44-74	74-44	DS07052	92E1	92E2	SAME CLONE AS DS06956
62-77	77-62	DS07358	92E1	92E2	ALSO CC LIGHT
42-5	5-42	DS00426	92E4	92E14	
45-84	84-45	DS08013	92E8,9	92E14	
	5-70	DS00454	92F1	92F2	
	6-27	DS00507	92F1	92F2	
81-37	Not present	DS09253	92F4	92F12	
	5-59	DS00443	93A2	93A4,5	MAYBE TO 93A1
64-80	80-64	DS07648	93A2	93A4,5	MAYBE TO 93A1 + CC LIGHT + ~100 SITES
45-18	18-45	DS01677	93A4,5		SAME CLONE AS DS01773
45-19	19-45	DS01773	93A4,5		SAME CLONE AS DS01677
20-19	19-20	DS01748	93A4,5	93B3	cDNA 9c5
25-24	24-25	DS02233	93B1,2	93B3	
47-65	65-47	DS06191	93B1,2	93B3	
25-89	89-25	DS08473	93B1,2	93B6	MAYBE TO 93B7
81-38	Not present	DS09254	93D1	93D5	
139-39	39-96	DS03744	93E1		
62-94	94-62	DS08990	93E1	93E2	
62-7	7-62	DS00638	93E3		
65-22	22-65	DS02081	93E9	93F4,5	MAYBE TO 93E8
137-8	8-94	DS00766	93F1	93F2	
	8-79	DS00751	93F2	93F4,5	MAYBE TO 93F1 + ~50 SECONDARY SITES
	9-17	DS00785	93F2	93F4,5	MAYBE TO 93F1 + 20-30 SECONDARY SITES
138-10	10-95	DS00959	93F2	93F4,5	MAYBE TO 93F1
47-12	12-47	DS01103	93F2	93F4,5	MAYBE TO 93F1 + ~25 SECONDARY SITES
43-43	43-43	DS04075	93F2	93F4,5	MAYBE TO 93F1
131-15	Not present	DS09327	93F4,5	93F8	
79-2	2-79	DS00175	93F7	93F9	MAYBE TO 93F6 &/OR 93F10
62-17	17-62	DS01598	93F9	93F10	
46-95	95-46	DS09070	93F9	93F10	
63-95	95-63	DS09087	93F9	93F10	
	4-32	DS00320	94B1	94B2	
46-12	12-46	DS01102	94B1	94B2	
90-21	21-89	DS02009	94B1	94B2	
135-16	16-92	DS01532	94B2	94B4	MAYBE TO 94B1 &/OR 94B5
25-25	25-25	DS02329	94B2	94B4	MAYBE TO 94B1 &/OR 94B5

CAMBR	BERK	DS No.	BEGIN	END	COMMENTS
41-56	56-41	DS05321	94B2	94B4	MAYBE TO 96B1 &/OR 96B5
43-32	32-43	DS03019	94C1,2	94C4,5	
	4-39	DS00327	94C4,5	94C9	
	1-17	DS00017	94D2	94D3	MAYBE TO 94D1 &/OR 94D4
60-2	2-60	DS00156	94D2	94D3	MAYBE TO 94D1 &/OR 94D4
139-8	8-96	DS00768	94D2	94D3	MAYBE TO 94D1 &/OR 94D4; SAME AS DS01536
139-16	16-96	DS01536	94D2	94D3	MAYBE TO 94D1 &/OR 94D4; SAME AS DS00768
135-36	36-92	DS03452	94D2	94D3	MAYBE TO 94D1 &/OR 94D4
62-62	62-62	DS05918	94D2	94D3	MAYBE TO 94D1 &/OR 94D4; ALSO 84A1;84A2 (LIGHT)
48-64	64-48	DS06096	94D2	94D3	MAYBE TO 94D1 &/OR 94D4
90-14	14-89	DS01337	94D3	94D4	
50-61	61-50	DS05810	94D3	94D4	
62-31	31-62	DS02942	94D4	94D10,11	MAYBE TO 94D3
49-13	13-49	DS01201	94D5	94D10,11	
	3-38	DS00230	94E1	94E3	MAYBE TO 94E4
50-86	86-50	DS08210	94E1	94E8	
45-92	92-45	DS08781	94E1	94E8	
24-35	35-24	DS03288	94E9,10	94F1,2	
50-89	89-50	DS08498	94E9,10	94F1,2	
	5-22	DS00406	94F1,2	94F6	
90-65	65-89	DS06233	94F3	94F6	
41-54	54-41	DS05129	94F6		
	1-14	DS00014	95A1	95A2	
25-66	66-25	DS06265	95A1	95A2	
	7-52	DS00628	95A2	95A4,5	MAYBE TO 95A1 + CC MODERATE
46-17	17-46	DS01582	95A2	95A4,5	MAYBE TO 95A1
	4-78	DS00366	95A4,5		ALSO CC LIGHT
64-18	18-64	DS01696	95A4,5		
46-59	59-46	DS05614	95A4,5		
90-31	31-89	DS02969	95B1,2		
50-78	78-50	DS07442	95B8	95C4	MAYBE TO 95C5
	7-43	DS00619	95C1,2	95C4	MAYBE TO 95C5 + CC MODERATE + ~200 SITES
	6-71	DS00551	95C8	95D1,2	MAYBE TO 95C7
	4-57	DS00345	95D1,2		
45-26	26-45	DS02445	95D1,2	95D5,6	
44-7	7-44	DS00620	95E1,2	95E5,6	
24-70	70-24	DS06648	95E5,6	95F1	MAYBE TO 95F2 + ~20 SECONDARY SITES
38-2	2-38	DS00134	96A1	96A2	
56-2	2-56	DS00152	96A1	96A2	
138-13	13-95	DS01247	96A1	96A2	
64-14	14-64	DS01312	96A1	96A2	
43-45	45-43	DS04267	96A1	96A2	
45-79	79-45	DS07533	96A1	96A2	
131-1	Not present	DS09313	96A1	96A2	
59-1	1-59	DS00059	96A3	96A9,10	
	5-6	DS00390	96A17	96A18	
24-81	81-24	DS07704	96A18	96A19	MAYBE TO 96A17 &/OR 96A20
131-44	Not present	DS09356	96A18	96A19	MAYBE TO 96A17 &/OR 96A20
	7-76	DS00652	96B1	96B2	
139-80	80-96	DS07680	96B2	96B6,7	MAYBE TO 96B1
	1-22	DS00022	96B10	96B11	ALSO CC LIGHT AND SOME SECONDARY SITES
44-55	55-44	DS05228	96B10	96B11	
50-66	66-50	DS06290	96B10	96B11	

CAMBR	BERK	DS No.	BEGIN	END	COMMENTS
	1-33	DS00033	96B11	96B14	MAYBE TO 96B10
138-26	26-95	DS02495	96B13	96C1	MAYBE TO 96C2
49-80	80-49	DS07633	96B20	96B21	MAYBE TO 96B19
24-57	57-24	DS05400	96C1	96C2	
43-63	63-43	DS05995	96C1	96C2	SAME AS DS06571
47-67	67-47	DS06383	96C1	96C2	
43-69	69-43	DS06571	96C1	96C2	SAME AS DS05995
135-95	95-92	DS09116	96C1	96C2	
48-96	96-48	DS09168	96C1	96C2	
81-23	Not present	DS09239	96C1	96C2	
45-27	27-45	DS02541	96C8	96C9	
	4-55	DS00343	96D1	96D2	ALSO CC LIGHT + ~25 VERY LIGHT SITES
48-63	63-48	DS06000	96D1	96D2	
131-4	Not present	DS09316	96D1	96D2	CC LIGHT TO MODERATE + ~50 SECONDARY SITES
	5-39	DS00423	96E2	96E3	MAYBE TO 96E1 &/OR 96E4
	7-31	DS00607	96E2	96E3	MAYBE TO 96E1 &/OR 96E4; SAME AS DS00608
	7-32	DS00608	96E2	96E3	MAYBE TO 96E1 &/OR 96E4; SAME AS DS00607
47-14	14-47	DS01295	96E2	96E3	MAYBE TO 96E1 &/OR 96E4
48-23	23-48	DS02160	96E2	96E3	MAYBE TO 96E1 &/OR 96E4
139-66	66-96	DS06336	96E2	96E3	MAYBE TO 96E1 AND/OR 96E4; =DS06528,DS06720
139-68	68-96	DS06528	96E2	96E3	MAYBE TO 96E1 AND/OR 96E4; =DS06336, DS06720
139-70	70-96	DS06720	96E2	96E3	MAYBE TO 96E1 AND/OR 96E4; =DS06336, DS06528
137-90	90-94	DS08638	96E2	96E3	MAYBE TO 96E1 &/OR 96E4
47-42	42-47	DS03983	96E5	96E12	
80-1	1-80	DS00080	96F1	96F2	
43-19	19-43	DS01771	96F1	96F2	
62-55	55-62	DS05246	96F1	96F2	
45-13	13-45	DS01197	96F1	96F3	MAYBE TO 96F4
	9-59	DS00827	96F3	96F4	ALSO SOME WEAK SECONDARY SITES
47-19	19-47	DS01775	96F3	96F4	ALSO CC LIGHT + ~100 SECONDARY SITES
49-26	26-49	DS02449	96F3	96F4	
50-26	26-50	DS02450	96F3	96F4	
50-47	47-50	DS04466	96F3	96F4	
139-15	15-96	DS01440	96F8	96F9	
64-30	30-64	DS02848	96F9	96F10,11	MAYBE TO 96F8
42-90	90-42	DS08586	96F9	96F10,11	MAYBE INTO 96F8 & LITE CC & ~100 SECONDARY SITES
42-91	91-42	DS08682	96F9	96F10,11	SAME AS DS08586
43-48	48-43	DS04555	97A6	97A7	MAYBE TO 97A5 &/OR 97A8 & ~25 OTHER SITES
24-69	69-24	DS06552	97A6	97A7	MAYBE TO 97A5 &/OR 97A8 + 5-6 AREAS WEAK SIGNAL
135-80	80-92	DS07676	97A6	97A7	MAYBE TO 97A5 &/OR 97A8
44-12	12-44	DS01100	97A9	97A10	
136-1	1-93	DS00093	97B1	97B2	
	9-1	DS00769	97B1	97B2	
42-66	66-42	DS06282	97B1	97B2	PRIMARY SITE
44-14	14-44	DS01292	97B4	97B10?	MAYBE TO 97B3 & ~100 OTHER SITES; THE ? IS SORSA'S
63-39	39-63	DS03711	97B4	97B10?	? DESIGNATION FROM SORSA; MAYBE TO 97B3; CC LIGHT
63-68	68-63	DS06495	97B5	97B10?	97B10? DESIGNATION IS SORSA'S
45-70	70-45	DS06669	97B6,7		
90-88	88-89	DS08441	97B6,7		
64-94	94-64	DS08992	97B6,7	97B10	
46-3	3-46	DS00238	97C1	97C3	
	7-36	DS00612	97C1	97C3	
43-95	95-43	DS09067	97C1	97D2	

CAMBR	BERK	DS No.	BEGIN	END	COMMENTS
45-44	44-45	DS04173	97C4	97C5	
44-32	32-44	DS03020	97D1	97D2	
46-48	48-46	DS04558	97D1	97D2	
42-51	51-42	DS04842	97D1	97D2	
25-61	61-25	DS05785	97D1	97D2	
46-83	83-46	DS07918	97D1	97D2	
67-2	2-67	DS00163	97D1	97D8,9	
139-40	40-96	DS03840	97D8,9		
45-41	41-45	DS03885	97E1	97E2	
42-45	45-42	DS04266	97E1	97E2	
131-64	Not present	DS09376	97E1	97E2	
139-84	84-96	DS08064	97E2	97E5	MAYBE TO 97E1 AND/OR 97E6
63-36	36-63	DS03423	97E6	97F1	MAYBE TO 97E5 &/OR 97F2
44-77	77-44	DS07340	97F2		
46-28	28-46	DS02638	97F6	97F11	
49-67	67-49	DS06385	98A1	98A2	
	3-90	DS00282	98A2	98A3	MAYBE TO 98A1
47-20	20-47	DS01871	98A2	98A3	MAYBE TO 98A1
	6-38	DS00518	98A2	98A4	MAYBE TO 98A1
48-25	25-48	DS02352	98A2	98A4	MAYBE TO 98A1
25-55	55-25	DS05209	98A2	98A4	MAYBE TO 98A1
81-35	Not present	DS09251	98A9	98A12	
	2-20	DS00116	98A11	98A12	
	9-87	DS00855	98A11	98A12	
139-19	19-96	DS01824	98A11	98A12	
46-14	14-46	DS01294	98B1	98B2	
24-19	19-24	DS01752	98B1	98B2	
	1-1	DS00001	98C1	98C2	
	3-88	DS00280	98C1	98C2	
	7-51	DS00627	98C1	98C2	ALSO SEVERAL SECONDARY SITES
46-66	66-46	DS06286	98C1	98C2	
50-76	76-50	DS07250	98C1	98C2	
66-2	2-66	DS00162	98D1	98D2	
	4-5	DS00293	98D1	98D2	
90-17	17-89	DS01625	98D1	98D2	
42-22	22-42	DS02058	98D1	98D2	
64-43	43-64	DS04096	98D1	98D2	
	4-88	DS00376	98D2	98D4	MAYBE TO 98D1 &/OR 98D5
65-37	37-65	DS03521	98D3	98D7	ALSO 63C5 LIGHT
	6-68	DS00548	98E1	98E2	ALSO CC MODERATE + ~50 SECONDARY SITES
44-48	48-44	DS04556	98E1	98E2	
	3-25	DS00217	98E3	98E6	
48-34	34-48	DS03216	98E3	98E6	
137-32	32-94	DS03070	98F1	98F2	
47-49	49-47	DS04655	98F1	98F2	
139-79	79-96	DS07584	98F1	98F2	
24-80	80-24	DS07608	98F3	98F6,7	
	8-71	DS00743	98F6,7		
44-57	57-44	DS05420	98F6,7		
43-65	65-43	DS06187	98F6,7		
131-81	Not present	DS09393	98F6,7		
136-91	91-93	DS08733	98F6,7	98F11,12	
47-62	62-47	DS05903	99A1,2	99A5,6	

CAMBR	BERK	DS No.	BEGIN	END	COMMENTS
63-25	25-63	DS02367	99A5,6		ALSO SOME SECONDARY SITES
139-88	88-96	DS08448	99A5,6		
81-4	Not present	DS09220	99A5,6	99A7,8	
139-4	4-96	DS00384	99A7,8		
44-37	37-44	DS03500	99A7,8		
	1-2	DS00002	99A9	99A11	
	2-29	DS00125	99B1	99B2	
137-7	7-94	DS00670	99B1	99B2	
137-27	27-94	DS02590	99B1	99B2	
137-47	47-94	DS04510	99B1	99B2	
138-43	43-95	DS04127	99B2	99B4	MAYBE TO 99B1 &/OR 99B5 + ~50 SECONDARY SITES
137-76	76-94	DS07294	99B2	99B4	MAYBE TO 99B1 &/0R 99B5 + ~25 SECONDARY SITES
	4-34	DS00322	99B4	99B5	
	9-66	DS00834	99B4	99B5	
42-31	31-42	DS02922	99B5		
136-77	77-93	DS07389	99B5		
49-95	95-49	DS09073	99B5	99B10	
50-95	95-50	DS09074	99B5	99B9	MAYBE TO 99B4
65-23	23-65	DS02177	99C2	99D1,2	MAYBE TO 99C1
	3-10	DS00202	99D1,2		
63-21	21-63	DS01983	99D1,2		
58-2	2-58	DS00154	99D1,2		ALSO CC MODERATE + ~100 WEAK SECONDARY SITES
	1-26	DS00026	99D1,2	99D5,6	
48-92	92-48	DS08784	99D5,6	99E2	MAYBE TO 99E3
138-19	19-95	DS01823	99E1	99E2	MAYBE TO 99E3
90-40	40-89	DS03833	99E1	99E2	MAYBE TO 99E3
45-59	59-45	DS05613	99E1	99E2	MAYBE TO 99E3
41-64	64-41	DS06089	99E1	99E3	MAYBE TO 99E4
49-83	83-49	DS07921	99E4	99E5	
24-52	52-24	DS04920	99E4	99F1	MAYBE TO 99F2
90-39	39-89	DS03737	99F1	99F2	
139-86	86-96	DS08256	99F1	99F2	ALSO CC LITE + SOME OTHER LITE EUCHROM SITES
45-88	88-45	DS08397	99F1	99F2	
65-92	92-65	DS08801	99F2	99F6	MAYBE TO 99F1 &/OR 99F7 + CC LITE + ~100 SITES
90-36	36-89	DS03449	99F7	99F8,9	MAYBE TO 99F6
24-62	62-24	DS05880	99F7	99F8,9	MAYBE TO 99F6
63-66	66-63	DS06303	99F7	99F8,9	MAYBE TO 99F6
	3-95	DS00287	100A1	100A2	
	8-54	DS00726	100A1	100A2	
44-43	43-44	DS04076	100A1	100A2	
137-53	53-94	DS05086	100A1	100A2	
50-55	55-50	DS05234	100A1	100A2	SAME AS DS08882?
46-72	72-46	DS06862	100A1	100A2	
25-85	85-25	DS08089	100A1	100A2	ALSO ~50 SECONDARY SITES
50-93	93-50	DS08882	100A1	100A2	SAME AS DS05234?
48-28	28-48	DS02640	100A3	100A5	MAYBE TO 100A6 + CC LIGHT
	7-39	DS00615	100B1	100B2	
46-67	67-46	DS06382	100B1	100B2	
	4-9	DS00297	100B4	100B5	
	6-69	DS00549	100B4	100B5	ALSO CC LIGHT + SOME FEW LIGHT SITES
47-51	51-47	DS04847	100B4	100B5	
62-78	78-62	DS07454	100B4	100B5	
47-2	2-47	DS00143	100B6	100B9	ALSO 10-20 WEAK SECONDARY SITES

CAMBR	*BERK*	*DS No.*	*BEGIN*	*END*	*COMMENTS*
44-64	64-44	DS06092	100B8	100B9	
	7-65	DS00641	100C1,2		
57-1	1-57	DS00057	100C1,2		
139-7	7-96	DS00672	100C1,2		
	8-65	DS00737	100C1,2		
43-41	41-43	DS03883	100C1,2		
45-43	43-45	DS04077	100C1,2		
63-58	58-63	DS05535	100C1,2		
135-76	76-92	DS07292	100C1,2		
45-94	94-45	DS08973	100C1,2		ALSO ~50 SECONDARY SITES (NO LABELING IN CC)
135-34	34-92	DS03260	100C4,5		
64-48	48-64	DS04576	100C4,5		
136-84	84-93	DS08061	100C4,5		
	5-16	DS00400	100D1	100D2	ALSO CC LIGHT + SOME LIGHT SECONDARY SITES
63-42	42-63	DS03999	100D1	100D2	
43-59	59-43	DS05611	100D2	100D4	MAYBE TO 100D1
43-3	3-43	DS00235	100D2	100E1	MAYBE TO 100D1 &/OR 100E2
43-44	44-43	DS04171	100D3	100E1	
64-24	24-64	DS02272	100E1	100E3	
81-76	Not present	DS09292	100F1	100F2	
	9-63	DS00831	100F1	100F5	
	4-80	DS00368	100F2	100F4	MAYBE TO 100F1 &/OR 100F5
62-23	23-62	DS02174	100F2	100F4	MAYBE TO 100F1 &/OR 100F5
46-2	2-46	DS00142	100F3	100F4	MAYBE TO 100F5
48-37	37-48	DS03504	100F3	100F5	TIP OF 3R
65-52	52-65	DS04961	100F3	100F5	VERY TIP
136-56	56-93	DS05373	100F3	100F5	ALSO ~ 20 WEAK SECONDARY SITES
25-96	96-25	DS09145	100F3	100F5	ALSO CC LIGHT + SOME SECONDARY SITES
90-84	84-89	DS08057	102B1,2		
65-50	50-65	DS04769	102C4		LIGHT BAND AT 102C4 + CC VERY LIGHT
42-41	41-42	DS03882	102C4	102C5	LIGHT SPECKS IN CC & LABEL IN 41C
81-5	Not present	DS09221	102C4	102C5	SAME AS DS09222
81-6	Not present	DS09222	102C4	102C5	SPECKLED CC & SITE IN 41C; SAME AS DSO9221
92-1	1-91	DS00091	102D1,2		
63-61	61-63	DS05823	102F5	102F8	OTHER TIPS WITH LIGHT SIGNALS

INDEX

Page numbers in italics denote figures (f) and tables (t).

M

N

O

P

R

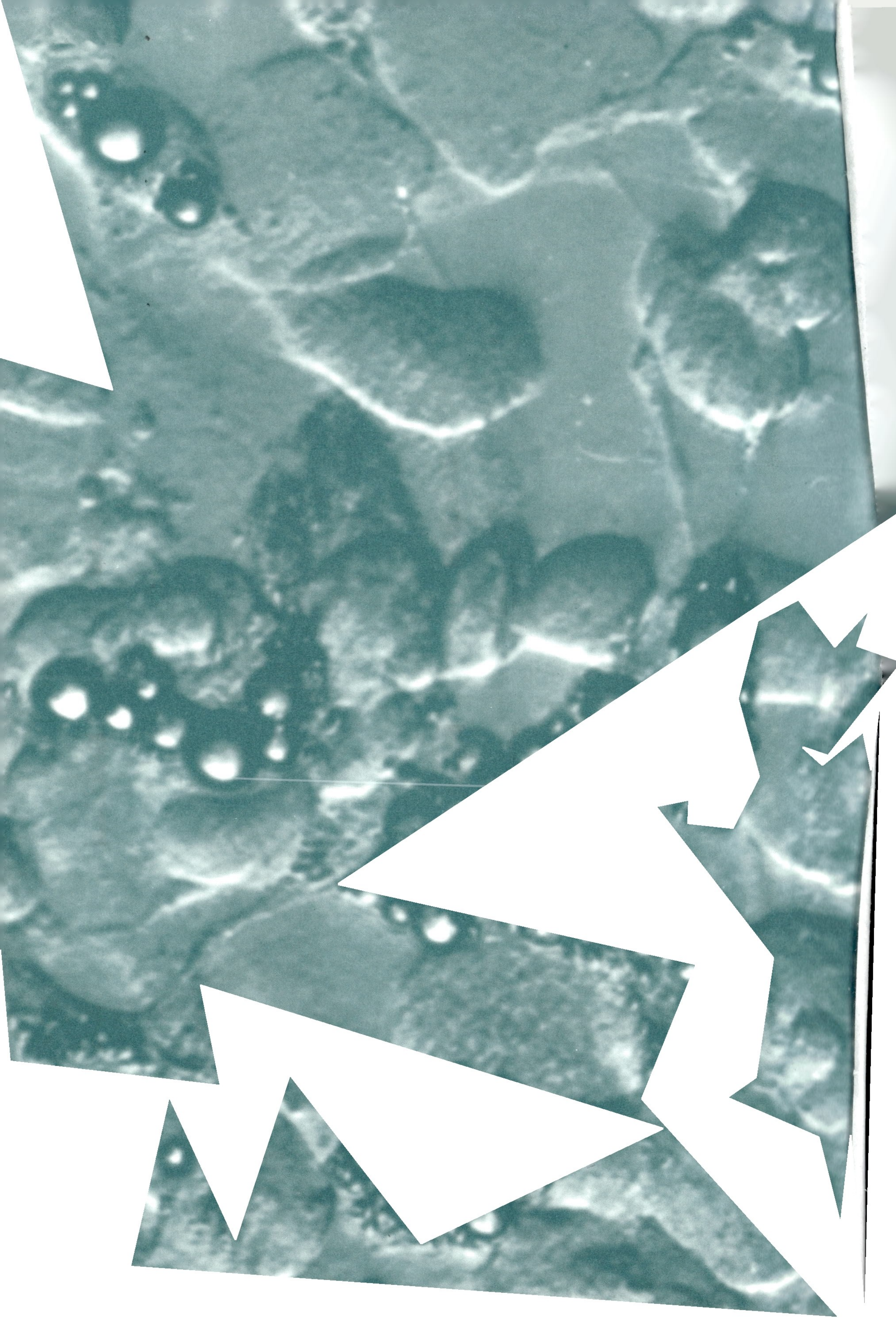

Medical Intelligence Unit

Available and Upcoming Titles

- Hyperacute Xenograft Rejection
  Jeffrey Platt, Duke University
- Chimerism and Tolerance
  Suzanne Ildstad, University of Pittsburgh
- Birth Control Vaccines
  G.P. Talwar and Raj Raghupathy, National Institute of Immunology-New Delhi and University of Kuwait
- Monoclonal Antibodies in Transplantation
  Lucienne Chatenoud, Hôpital Necker-Paris
- Therapeutic Applications of Oligonucleotides
  Stanley Crooke, ISIS Pharmaceuticals
- Cryopreserved Venous Allografts
  Kelvin G.M. Brockbank, CryoLife, Inc.
- Clinical Benefits of Leukpodepleted Blood Products
  Joseph Sweeney and Andrew Heaton, Miriam and Roger Williams Hospitals-Providence and Irwin Memorial Blood Center-San Francisco
- Delta Hepatitis Virus
  M. Dinter-Gottlieb, Drexel University
- Intima Formation in Blood Vessels: Spontaneous and Induced
  Mark M. Kockx, Algemeen Ziekenhuis Middelheim-Antwerpen
- Adult T Cell Leukemia and Related Diseases
  Takashi Uchiyama and Jungi Yodoi, University of Kyoto
- Development of Epstein-Barr Virus Vaccines
  Andrew Morgan, University of Bristol
- p53 Suppressor Gene
  Tapas Mukhopadhyay, Steven Maxwell and Jack A. Roth, University of Texas-MD Anderson Cancer Center
- Retinal Pigment Epithelium Transplantation
  Devjani Lahiri-Munir, University of Texas-Houston
- Minor Histocompatibility Antigens and Transplantation
  Craig V. Smith, University of Pittsburgh
- Familial Adenomatous Polyposis Coli and the APC Gene
  Joanna Groden, University of Cincinnati
- Cancer Cell Adhesion and Tumor Invasion
  Pnina Brodt, McGill University
- Constitutional Immunity to Infection
  Cees M. Verduin, David A. Watson, Jan Verhoef, Hans Van Dijk, University of Utrecht and North Dakota State University
- Nutritional and Metabolic Support in Critically Ill Patients
  Rifat Latifi, Yale University
- Nutritional Support in Gastrointestinal Failure
  Rifat Latifi and Stanley Dudrick, Yale University and University of Texas-Houston
- Septic Myocardiopathy: Molecular Mechanisms
  Karl Werdan and Günther Schlag, Ludwig-Maximilians-Universität-München and Ludwig-Boltzmann-Instituts für Experimentelle und Klinische Traumatologie
- The Molecular Genetics of Wilms Tumor
  Bryan R.G. Williams, Max Coppes and Christine Campbell, Cleveland Clinic and University of Calgary
- Endothelins
  David J. Webb and Gillian Gray, University of Edinburgh
- Nutritional and Metabolic Support in Cancer, Transplant and Immunocompromised Patients
  Rifat Latifi, Yale University
- Antibody-Mediated Graft Rejection
  J. Andrew Bradley and Eleanor Bolton, University of Glasgow
- Liposomes in Cancer Chemotherapy
  Steven Sugarman, University of Texas-MD Anderson
- Molecular Basis of Human Hypertension
  Florent Soubrier, Collége de France-Paris
- Endocardial Endothelium: Control of Cardiac Performance
  Stanislas U. Sys and Dirk L. Brutsaert, Universiteit Antwerpen
- Endoluminal Stent Grafts for the Treatment of Vascular Diseases
  Michael L. Marin, Albert Einstein College of Medicine
- B Cells and Autoimmunity
  Christian Boitard, Hôpital Necker-Paris
- Immunity to Mycobacteria
  Ian Orme, Colorado State University
- Hepatic Stem Cells and the Origin of Hepatic Carcinoma
  Stewart Sell, University of Texas-Houston
- HLA and Maternal-Fetal Recognition
  Joan S. Hunt, University of Kansas
- Molecular Mechanisms of Alloreactivity
  Robert L. Kirkman, Harvard University
- Ovarian Autoimmunity
  Roy Moncayo and Helga E. Moncayo, University of Innsbruck
- Immunology of Pregnancy Maintenance
  Joe Hill and Peter Johnson, Harvard University and University of Liverpool
- Protein and Amino Acid Metabolism in Cancer
  Peter W.T. Pisters and Murray Brennan, Sloan-Kettering Memorial Cancer Center
- Cytokines and Hemorrhagic Shock
  Eric J. DeMaria, Medical College of Virginia
- Cytokines in Inflammatory Bowel Disease
  Claudio Fiocchi, Case Western Reserve University
- T Cell Vaccination and Autoimmune Disease
  Jingwu Zhang, Willems Institut-Belgium
- Immune Privilege
  J. Wayne Streilein, Luke Jiang and Bruce Ksander, Schepens Eye Research Institute-Boston
- The Pathophysiology of Sepsis and Multi-Organ Failure
  Mitchell Fink, Harvard University
- Bone Metastasis
  F. William Orr, McMaster University
- Novel Regional Therapies for Liver Tumors
  Seiji Kawasaki and Masatoshi Makuuchi, Shinshu University
- Molecular Basis for the Action of Somatostatin
  Miguel J.M. Lewin, INSERM-Paris
- Growth Hormone in Critical Illness
  Michael Torosian, University of Pennsylvania
- Molecular Biology of Aneurysms
  Richard R. Keen, Northwestern University
- Strategies in Malaria Vaccine Design
  F.E.G. Cox, King's College London
- Chimeric Proteins and Fibrinolysis
  Christoph Bode, Marschall Runge and Edgar Haber, University of Heidelberg, University of Texas-Galveston and Harvard University

Available Free Titles

Please check three titles in order of preference.
Your request will be filled based on availability. Thank you.

- ❐ Water Channels
  Alan Verkman,
  University of California-San Francisco
- ❐ The Na,K-ATPase:
  Structure-Function Relationship
  J.-D. Horisberger, University of Lausanne
- ❐ Intrathymic Development of T Cells
  J. Nikolic-Zugic,
  Memorial Sloan-Kettering Cancer Center
- ❐ Cyclic GMP
  Thomas Lincoln, University of Alabama
- ❐ Primordial VRM System and the Evolution of Vertebrate Immunity
  John Stewart, Institut Pasteur-Paris
- ❐ Thyroid Hormone Regulation of Gene Expression
  Graham R. Williams, University of Birmingham
- ❐ Mechanisms of Immunological Self Tolerance
  Guido Kroemer, CNRS Génétique Moléculaire et Biologie du Développement-Villejuif
- ❐ The Costimulatory Pathway for T Cell Responses
  Yang Liu, New York University
- ❐ Molecular Genetics of Drosophila Oogenesis
  Paul F. Lasko, McGill University
- ❐ Mechanism of Steroid Hormone Regulation of Gene Transcription
  M.-J. Tsai & Bert W. O'Malley, Baylor University
- ❐ Liver Gene Expression
  François Tronche & Moshe Yaniv,
  Institut Pasteur-Paris
- ❐ RNA Polymerase III Transcription
  R.J. White, University of Cambridge
- ❐ src Family of Tyrosine Kinases in Leukocytes
  Tomas Mustelin, La Jolla Institute
- ❐ MHC Antigens and NK Cells
  Rafael Solana & Jose Peña,
  University of Córdoba
- ❐ Kinetic Modeling of Gene Expression
  James L. Hargrove, University of Georgia
- ❐ PCR and the Analysis of the T Cell Receptor Repertoire
  Jorge Oksenberg, Michael Panzara & Lawrence Steinman, Stanford University
- ❐ Myointimal Hyperplasia
  Philip Dobrin, Loyola University
- ❐ Transgenic Mice as an In Vivo Model of Self-Reactivity
  David Ferrick & Lisa DiMolfetto-Landon, University of California-Davis and Pamela Ohashi, Ontario Cancer Institute
- ❐ Cytogenetics of Bone and Soft Tissue Tumors
  Avery A. Sandberg, Genetrix & Julia A. Bridge ,
  University of Nebraska
- ❐ The Th1-Th2 Paradigm and Transplantation
  Robin Lowry, Emory University
- ❐ Phagocyte Production and Function Following Thermal Injury
  Verlyn Peterson & Daniel R. Ambruso,
  University of Colorado
- ❐ Human T Lymphocyte Activation Deficiencies
  José Regueiro, Carlos Rodríguez-Gallego and Antonio Arnaiz-Villena,
  Hospital 12 de Octubre-Madrid
- ❐ Monoclonal Antibody in Detection and Treatment of Colon Cancer
  Edward W. Martin, Jr., Ohio State University
- ❐ Enteric Physiology of the Transplanted Intestine
  Michael Sarr & Nadey S. Hakim, Mayo Clinic
- ❐ Artificial Chordae in Mitral Valve Surgery
  Claudio Zussa, S. Maria dei Battuti Hospital-Treviso
- ❐ Injury and Tumor Implantation
  Satya Murthy & Edward Scanlon,
  Northwestern University
- ❐ Support of the Acutely Failing Liver
  A.A. Demetriou, Cedars-Sinai
- ❐ Reactive Metabolites of Oxygen and Nitrogen in Biology and Medicine
  Matthew Grisham, Louisiana State-Shreveport
- ❐ Biology of Lung Cancer
  Adi Gazdar & Paul Carbone,
  Southwestern Medical Center
- ❐ Quantitative Measurement of Venous Incompetence
  Paul S. van Bemmelen, Southern Illinois University and John J. Bergan, Scripps Memorial Hospital
- ❐ Adhesion Molecules in Organ Transplants
  Gustav Steinhoff, University of Kiel
- ❐ Purging in Bone Marrow Transplantation
  Subhash C. Gulati,
  Memorial Sloan-Kettering Cancer Center
- ❐ Trauma 2000: Strategies for the New Millennium
  David J. Dries & Richard L. Gamelli,
  Loyola University